C0 CEM 590

ECONOMIC DECLINE AND NATIONALISM IN THE BALKANS

Also by Milica Zarkovic Bookman

THE ECONOMICS OF SECESSION

THE POLITICAL ECONOMY OF DISCONTINUOUS DEVELOPMENT

ISSUES IN INDIAN AGRICULTURAL DEVELOPMENT
(*writing as Milica Zarkovic*)

ECONOMIC DECLINE AND NATIONALISM IN THE BALKANS

Milica Zarkovic Bookman

St. Martin's Press
New York

 For information, write:
Scholarly & Reference Division
St. Martin's Press, 175 Fifth Avenue,
New York, N.Y. 10010

First published in the United States of America 1994
Printed in the United States of America
Book design by Acme Art, Inc.

ISBN 0-312-09999-1

Library of Congress Cataloging-in-Publication Data

Bookman, Milica Zarkovic.
Economic decline and nationalism in the Balkans / Milica Zarkovic Bookman.
p. cm.
Includes bibliographical references (p.) and index.
ISBN 0-312-09999-1
1. Balkan Peninsula—Economic conditions. 2. Nationalism—Balkan Peninsula. 3. Balkan Peninsula—Social conditions. 4. Balkan Peninsula—Politics and government—20th century. I. Title.
HC401.B66 1994
330.947'4—dc20 93-47026
CIP

To Marco, Karla, Cristina and Aleksandra,
In hope that their offspring will, one day, play along Ulica Prijeko

CONTENTS

LIST OF MAPS AND TABLES

ACKNOWLEDGMENTS

I am indebted to numerous people who have encouraged this research and provided assistance in all its phases. A large part of the research and writing of this book occurred during the year that I commuted between Miami and Philadelphia. During this year, I could allocate undisturbed time every week to writing. For enabling this schedule, and believing that I could survive it, I am grateful to George Prendergast and Vincent McCarthy.

Some research for this book was conducted in the Balkans. I am grateful for the support from St. Joseph's University, which enabled me to study in Greece during the summer of 1993. Special thanks go to Professor Dragas Denkovic and Alexis Mitropulos for helping with arrangements there, as well as to Louka Katseli for her time and assistance. I would also like to thank Tomislav Popovic and Miladin Kovacevic for the help, as well as the access to data and unpublished materials, that they extended to me. Richard Bookman prepared the maps, for which I am grateful.

A special word of thanks to my editor, Simon Winder, who has always been helpful and understanding, despite his apprehension about my enormous tables.

Finally, I want to thank Richard, Karla and Aleksandra for providing the balance in my life that brought peace in times of stress and stability in times of turbulence.

1

Nationalist Bankruptcy in the Balkans

"Someone who went to sleep in 1939 and woke up in 1992 might think little has changed."

Andrei Codrescu, writing about minority rights in Europe[1]

"History has not ended; only that great master of discipline, the cold war, has taken a bow. The little wars have just begun."

Josef Joffe[2]

"Nationalism appears to neutralize that part of the mind which is able to fathom complex equations. Instead, action is motivated by a single Leninist principle: 'Those who are not for us, are against us.' "

Misha Glenny[3]

In the 1990s, two major forces are tearing up states across the globe and are causing chaos in the post-cold war order. These are nationalism and economic decline. The former, often but not entirely the result of decades of suppression, has manifested itself in outbursts of xenophobic intolerance between ethnic, religious and historical enemies. The latter has been characterized by decreases in output and employment, a deterioration in the trade balance, galloping inflation, brain drain and the proliferation of underground economies. In some unfortunate countries, the 1990s have witnessed a convergence of both nationalism and economic decline, resulting in a virtual breakdown of the social and economic fabric. It is in the Balkans, especially in the former Yugoslav republics, that these pressures are presently the most concentrated. Former Yugoslavia is at war over ethnic intolerance and economic deprivation; Romania is buckling under minority and economic pressures; Albania and Bulgaria are unable to control their economic decline while their ethnic concerns spill outside of their boundaries, notably into Kosovo and Macedonia respectively; and Greece is reeling under popular displeasure with the austerity measures imposed by the Mitsotakis government and is thus susceptible to nationalist fever centering on the questions of Macedonia and North Epirus. All Balkan economies are adversely affected by the sanctions imposed on the new Yugoslavia, and all Balkan peoples are in fear of the Yugoslav war spreading to their lands. In the 1990s, no Balkan country is going unscathed, and on several occasions it has seemed that, one more time in this century, a larger world conflict might be sparked in Sarajevo.

The Balkans are back on the center stage of world events and the attention of the media, policy makers and academics has once again in this century turned to this region, not only because of the confluence of nationalism and economic decline, but for other reasons as well: First, the civil war in former Yugoslavia has indicated with clarity the devastation that uncontrolled nationalism can bring about and raises the possibility of the conflict spreading into neighboring states and involving Albania, Greece, Bulgaria and Turkey, for reasons similar to those that led to their involvement in the past two centuries. Indeed, Glenny has already called the Yugoslav war the Third Balkan War.[4] Second, the Balkans are in a part of the world that is struggling with the transition from communism to capitalism, a transition which itself is under the microscope. Indeed, because four of the former Soviet Bloc states are part of the Balkans, observers of the breakdown of communism are likely to focus on this region. Third, the end of the cold war and the demise of a bipolar political world have resulted in the global dominance of a single superpower, the United States. The postwar political arena is in a state of flux, as the United States is groping for a way to define its role as the sole superpower and the European Community is unsure how to proceed in the post-Maastrich era. But at the regional level, there is competition in the establishment of a new balance of power. In the Balkans, this has translated into the surge in influence of Germany and Turkey as they attempt to reposition themselves within the region and find new roles. Indeed, unified Germany had by 1991 achieved its pre-World War II level of economic investment in the region, and its political involvement is clearly exemplified by its lead in the recognition of Yugoslav successor states. Turkey is also active in extending its influence economically in the form of aid and investment in Macedonia, Albania and Bulgaria and politically by offering strong support to the Bosnian Muslims. Thus, against the background of the vacuum created by the demise of the Soviet Union and the failure of the United States to define its new role in the world, Germany and Turkey are vying for influence in the Balkans.

This book is about the precipitous economic decline and rampant nationalism in the Balkans in the 1990s. The study asks the following three questions: First, what internal and external factors have deepened the economic crises in the Balkan states over the past three years? Second, what is the nature of nationalism in the Balkans? Third, what is the relationship between economic decline and the rise of nationalism? Chapter 1 describes recent events in the Balkans using the concept of "nationalist bankruptcy". In chapter 2, the ethnoterritorial pressures presently evident in the Balkans are discussed, and the principle of self-determination is explored within the Balkan context. Economic decline is studied in chapter 3, and the combined effect on economies of the transition to capitalism and the Yugoslav war is covered in chapter 4. In chapter 5, the ramifications of sanctions imposed on Yugoslavia are studied, while in chapter 6 the population movements across the Balkans are discussed. By evaluating the costs and benefits of fragmentation and integration in the Balkan states, chapter 7 explores the economic future of the Balkans. I argue that, given the nature of the economic and ethnic composition of the Balkan states, larger heterogeneous units are preferable to the proliferation of small nation-states.

NATIONALIST BANKRUPTCY IN THE LATE TWENTIETH CENTURY

Numerous societies are experiencing multidimensional crises in the post-cold war period. These crises, which manifest themselves in the unravelling social, economic and political fabric of states, are not confined to specific geographical regions, systems or levels of development: indeed, they are universal insofar as they cut across political and economic systems as well as across levels of development. In other words, crises are not limited to the postcommunist world (witness Greece and Italy), nor the less developed regions (witness Czechoslovakia and the Baltic states), nor only to federations (witness unitary states such as Romania and Sudan). The global extent of this turmoil was pointed out by Helman and Ratner, who claimed that "from Haiti in the Western Hemisphere to the remnants of Yugoslavia, from Somalia, Sudan and Liberia in Africa to Cambodia in Southeast Asia, a disturbing new phenomenon is emerging: the failed nation-state, utterly incapable of sustaining itself as a member of the international community...

those states descend into violence and anarchy, imperiling their own citizens and threatening their neighbors through refugee flows, political instability and random warfare."[5]

Common elements among many countries that have succumbed to instability and war are severe economic decline and rampant nationalism. While economic decline is easy to identify, as it is measured by a variety of economic indicators such as output and unemployment, nationalism is significantly more difficult to identify, measure and understand. According to Anthony Smith, nationalism is defined as "a doctrine of autonomy, unity and identity for a group whose members conceive it to be an actual or potential nation," where a nation is "a body of citizens bound by shared memories and a common culture, occupying a compact territory with a unified economy and identical rights and duties."[6] Culture, ethnicity and language are all embodied in this notion of nationalism, and as such are, according to Allcock, "portmanteau terms which can be used to accommodate virtually anything."[7] Indeed, issues of ethnicity, religion and language are often so interconnected that they blend with each other, making it impossible to quantify, in a scientific manner, how much each of these factors has contributed to the nationalist sentiment. Ethnicity contributes to nationalism when injustices are perceived to occur on an ethnic basis: the Ibos, Tibetans, Punjabis, Kashmiris, Albanians, et cetera all claim mistreatment on the basis of ethnicity. Religion is similar to ethnicity in its effect on nationalism insofar as it is a characteristic that sets people apart from outsiders and may be a cause of discrimination.[8] Nationalism might also have a linguistic component, such as when a population attempts to protect and preserve a language and script in the face of pressure from a linguistically dominant majority, as in the case of the Catalans in Spain, the Quebecois in Canada or the Tamils in Sri Lanka.[9]

When nationalist pressures combine with economic decline, it creates internal conditions that lead to the disintegration of the fabric that defines the economic, political and social system. One element of the explosive mixture of economic decline and nationalism is the emergence of "nationalist bankruptcy." This term describes the condition of a society in which the nationalist policies and demands of an ethnic or religious group become destructive not only for the society it is a part of, but also for the group itself. It describes a situation in which the mushrooming demands by ethnic groups paralyze the functioning of economic and political systems. At such a time, ethnically based territorial demands come to the forefront, and territorial fragmentation according to ethnic group is perceived as a solution to economic stagnation, political repression and social malaise. Thus nationalist bankruptcy is a state of affairs that comes to pass when other, more tolerant and moderate, options for ethnic cooperation are exhausted and ethnic purity is perceived to be a panacea. Nationalist bankruptcy occurs when ethnic groups, overwhelmed by economic hardship and frightened of their changing position relative to other ethnic groups, engage in pursuit of ethnic purity as the ultimate act of triumph and desperation. When the right conditions exist in the international environment, then such nationalist bankruptcy brings a society to the verge of collapse and produces a "national sclerosis." Thus the term "nationalist bankruptcy" connotes the decline of institutions due to the mix of belligerent interethnic relations with a collapsing economy. In most cases in history, such a potentially explosive mixture has resulted in wars, large scale migrations and other forms of human suffering.

Nationalist bankruptcy has as a precondition economic decline and a decrease in the standard of living. It then manifests itself in various economic, social and political ways, aided by the right conditions in the outside world. Some of the components of nationalist bankruptcy listed below are simultaneously the source, the concomitant and the ramification of nationalist bankruptcy, since the components all reinforce one another as they spiral out of control. They are described in the following categories: economic components, ethnicity and nationalism, political components and the external environment.

Economic Components

Economic Decline

The following manifestations of economic stagnation and decline tend to precede and then fuel nationalist bankruptcy: First, there is a decline in output, as the supply of goods and services falls over time. This translates into a decline in income, both at the personal and national levels. Both of these result in a decrease in the standard of living; the former directly affects the household, while the latter affects it indirectly over the long term as social programs and benefits are reduced. Furthermore, they bring about a decrease in demand, fueling the recessionary cycle. The decrease in output may be preceded or followed by a decline in productivity of all factors of production. Moreover, increasing inflation and decreasing employment make domestic policy difficult to formulate and carry out. Domestically, different regions and different income groups are affected in different ways, thus increasing inequality among the population. With respect to international economic relations, economic decline usually includes a deteriorating balance of payments, a stagnation in foreign investment and inflow of capital and disarray in international financial markets. The following micro conditions are also present: capital shortage for firms; high rate of firm bankruptcies; and the inability of the economic system to maintain a sufficient supply of consumer goods. In addition, there is a seeming lack of effective monetary and fiscal policy to control events and reverse the trend of economic decline. Sometimes this leads to economic paralysis as institutions fail to perform and require massive intervention to prop them up.

Economic Decline and Nationalism

In multiethnic societies, especially those in which substate divisions are drawn according to ethnic, religious or linguistic lines, regional economic competition is easily interpreted as interethnic competition. This is common in states such as India, Yugoslavia, the Soviet Union, Czechoslovakia and perhaps even Canada and Spain. Indeed, Punjabis in India perceive that in the competition for scarce resources at the central level, their state of Punjab, and by extension their ethnic group, is discriminated against. This same perception of regional economic issues underlies the efforts at secession of the Slovenes.

When economic conditions deteriorate, competition becomes more ferocious and fuels nationalist ideology. The above macro and micro problems may become exacerbated as interethnic bickering imperils economic functioning and paralyzes economic institutions, contributing to further macro and micro failures. The critical consideration among populations is the perception of economic injustice, in regard to both objective macro conditions, such as poverty, and policy aimed at rectifying those conditions, resulting in the following: above-average contribution to the national budget, insufficient benefit from the national budget, unfavorable terms of trade resulting from price manipulation, unfavorable regulation pertaining to investment and foreign inflows of resources, and so on. It is clear that perceptions of economic exploitation may be experienced by populations in regions that are more *or* less developed relative to the nation, as is evident in Italy (Lombardy as well as the Mezzogiorno), India (Punjab as well as Kashmir), Yugoslavia (Slovenia as well as Macedonia) and in the former USSR (Lithuania as well as Turkmenia). The high-income, subnational regions such as Lombardy, Punjab, Slovenia and Lithuania experienced tax revolts, reflecting dissatisfaction with what they perceived to be an unfair drainage of their resources, while the less developed regions lobbied for increased "spread effects" of national development, as well as a change in the redistributive policy. Perceptions of economic injustice influence the valuation of relative costs and benefits of belonging to a national union, and when costs outweigh benefits, economic factors are then mingled with ethnic, religious or cultural factors to form a set of demands that may include leaving the union.

In the absence of interethnic animosities, or under conditions of ethnic homogeneity, such interregional competition may simply be regionalism. This is especially evident in states that contain marked regional income inequalities, a phenomenon identified in the literature as uneven economic development,[10] discontinuous development,[11] relative deprivation,[12] internal colonialism[13] and differential modernization.[14] It produces dissatisfaction in the better-off regions because they perceive the central government as redistributing in favor of the less developed regions. In some regions, the result is limited to a simple tax revolt; in others, a major social and political movement grows, as it has in Italy with the Northern Leagues. When ethnic or religious distinctions coincide with regional administrative boundaries, then the soil is fertile for the emergence of a nationalist and perhaps even separatist movement. There are numerous countries in which large differences in levels of development exist, such as the United States and Italy, yet there exists no nationalism because of the lack of association between ethnic identity and the region. On the contrary, in the Balkans, different levels of development have become an ethnic issue, as they are associated with nationality. Simmie and Dekleva claim that the reason for the chaos in Yugoslavia, in addition to nationalism, is the distribution of economic resources. Indeed, they stress that the basis of the current conflicts is the "economic wars between the richer northern republics and the poorer southern ones."[15] The region of Yugoslavia inhabited almost exclusively by the Slovenes, is said to be 7.5 times wealthier than Kosovo.[16] Similarly, Transylvania, the more developed and industrialized region of Romania, is inhabited mostly by Hungarians. Thus economic development becomes an ethnic issue, creating a link between nationalism and economics.

Is there a causal link between economic decline and nationalism?[17] Most literature discussing this link has focused on economic development rather than decline. A large body of this literature suggests that nationalism tends to decrease with modernization and thus, by implication, that it would increase with economic decline. Indeed, it was accepted among scholars and policy makers that economic development, as exemplified by capitalist industrialization, would have the effect of reducing group identities and therefore would dissipate separatist tendencies.[18] According to Deutsch and

Huntington, with modernization, ethnic groups tend to assimilate,[19] thereby decreasing the possibility of ethnically based demands. Some claim that modernization raises the educational and cultural levels of the population, thus making people think and act as members of civic societies not based on ethnic communities.[20] However, this view was refuted by the proliferation of ethnic self-assertion activity of the mid-1960s in Western European industrialized countries such as France, Britain and Belgium. Thus, there emerged another group of theories that claimed that industrial capitalism was in fact conducive to ethnic protest and that the political arena that accompanies modernization provides a vehicle for nationalist self-expression.[21] Connor suggested that with modernization and the increasing ability of a population to communicate, the tendency is to increase self-awareness, and the distinction of the ethnic group is enforced.[22] The most common description of the relationship between ethnic demands and economic development is an inverse one: namely, the greater the underdevelopment, deterioration and stagnation of a regional economy, the greater the efforts of the ethnic group to differentiate itself from the union (Hroch found this in nineteenth-century Europe, Michneck in contemporary USSR, Birch in Bangladesh[23]). Indeed, Drake claims that economic inequality "was a major underlying cause of the civil war in Sudan, and it had a definite role in the breakup of Pakistan into Pakistan and Bangladesh. Within Indonesia, too, several of the regional rebellions experienced since independence have had economic grievances at their root."[24] The direct relationship between economic development and regional/ethnic demands has rarely been drawn: Wallerstein is rare among scholars to argue that the wealthy regions, as a result of their wealth, are more likely to make extreme demands, such as secession.[25] This relationship has also been identified and elaborated upon by the author with a comparative study of several secessionist movements.[26]

While the above discussion has explored the effects of the economy on nationalism, one might instead explore the role nationalism plays on economic development. This was done by Gershenkron, who claims that nationalism enables a society "to break through the barriers of stagnation in a backward country, to ignite the imaginations of men, and to place their energies in the service of economic development." This view helps us understand his definition of nationalism as "an ideology of delayed industrialization."[27] Nationalism is further seen as a conduit for economic growth insofar as it enables, according to Geller, communication through a literate, educated culture, and serves as a bond among like peoples.[28]

Thus theory claims, and practice illustrates, that nationalism and ethnically based demands on the union can occur at all levels of development, including on one end of the spectrum Quebec, Lombardy and the Baltic states of the USSR, and at the other, the less developed Punjab, Basque provinces, Moldova and Bougainville.

Ethnicity and Nationalism

Nationalism and Territorial Goals

Nationalist bankruptcy is associated with a proliferation of expressions of nationalism that are ethnic in focus, exclusionary in nature and territorial in aspiration. That nationalism is ethnic implies that ethnicity becomes the chief defining characteristic and ethnic purity its chief goal, as opposed to independence or self-determination, both of which can be regional in nature. Self-determination was the goal of the nationalism that brought about democratic change in France preceding the French Revolution, as well as in the dominions of the Hapsburg and Ottoman Empires and in independence movements across the Third World. It is this type of nationalism that caused Rosa Luxemburg to view it as a progressive force. Zaslavsky also views nationalism in postcommunist societies as a positive force that helps shed vestiges of communism, bring about market reforms and ultimately lead to new forms of international integration: "The liberal-democratic potential of the new nationalism needs to be fully appreciated."[29]

The exclusionary element of nationalism refers to the way in which ethnicity is defined, namely by the exclusion of others. This takes the form of overt measures that exclude ethnic groups, such as language laws in Sri Lanka or the constitutionally defined lack of rights of nontitular ethnic groups, such as in Latvia or Croatia. Thus Hobsbawm argues that "the characteristic nationalist movements of the late twentieth century are essentially . . . divisive."[30] Talmon also perceives such exclusion as suppressing individual rights and initiative.[31] Very intense sentiment in favor of one's nation and against others is often born in part out of fear and results in aggressive behavior. This kind of xenophobia, in which members of an ethnic group are fearful and hostile of others, gives rise to the need for scapegoats. Most often this role is filled by other ethnic groups (for example, Gypsies or Jews) or institutions, as with the legacy of communism. Xenophobia is present in Germany in a mild form, in the Czech Republic it is stronger, while in former Yugoslavia it is strongest. In the last, there is an unwillingness to attempt multiethnic survival, as members of ethnic groups follow the policy of "cleanse or be cleansed."[32]

Finally, the territorial element of nationalist bankruptcy is crucial, since it entails the bestowing of ethnic rights by the creation of sovereignty over a territory. Territorial demands differentiate a nationalist movement from those that merely demand additional privilege within a given political system. These may simply include demands for increased favoritism by the center toward a region or a targeted segment of the population, or they may be demands for a dramatic change in the participation of a region in the central and state affairs. Alternatively, the demands may be such that nothing short of severance of preexisting economic and political ties with the center is acceptable. This demand, which is actually the only secessionist demand, is referred to by Leslie as the "we want out" demand,[33] while Bremmer modified Hirshman's concept and calls it the "exit option."[34] In addition to secession, ethnic demands may include the establishment of ethnic supremacy and sovereignty on a heretofore ethnically undefined territory, as is presently occurring in the struggle for ethnic domination by Bosnian Serbs, Bosnian Croats and Bosnian Muslims in Bosnia-Herzegovina.

Armed Struggle

Ethnic movements that are exclusionary in nature and have territorial goals are often characterized by the participants' willingness to take up armed struggle for the cause. In some cases, this involves localized guerrilla warfare, such as among the Karen in Myanmar or the terrorist activities of the Corsicans or the Quebecois during the 1970s. The Irish struggle for secession from Britain has been significantly more violent. In some cases, civil war has occurred over extreme, ethnically based territorial demands, such as in Croatia, Biafra and Katanga. Indeed, the conflict in the Congo pertaining to Katanga almost produced World War III, as international economic and geopolitical interests became threatened. Today the intense fighting in Angola, Afghanistan and former Yugoslavia are clear examples of the tenacity of ethnic causes and the relative ease with which they translate into armed struggle. This tenacity is evident in the longevity of interethnic warfare, exemplified clearly in the persistent struggle of the Palestinians, the Eritreans, the Kurds, the Southern Sudanese and the Northern Irish, all of whom have been at war for decades.

Large-Scale Population Movements

One of the defining characteristics of nationalist bankruptcy is the drive for ethnic purity of a territory. Such a drive is based in part on intolerance, but more often on fear and the perception of danger associated with other ethnic groups. The crucial element driving individuals in interethnic armed struggles is most often genuine fear for themselves, their families, their property and their ancestral ties, rather than territorial ambition.[35] The principal by-product of this quest for ethnic purity is the large-scale movement of populations, as refugees relocate out of areas where they are persecuted and into areas of relative safety where they are the majority.

These large-scale population movements are part of nationalist bankruptcy insofar as they are the result of efforts by one ethnic group to use demography to gain political power with respect to another group. The population shifts then have as their goal the immediate safety of the escaping ethnic group and the demographic control of the remaining ethnic group. The economic costs of the loss of labor incurred by the region being fled is tremendous, further aggravating the conditions that give rise and perpetuate nationalist bankruptcy.

Population movements due to forced relocations based on ethnicity do not occur in ethnically homogeneous states. Since there are very few of these, it is to be expected that ethnically motivated population movements will occur. Indeed, a study by Connor shows that in the late twentieth century, out of a total of 132 states, only twelve are ethnically homogeneous.[36] If these heterogeneous states have regional subdivisions that coincide with ethnic borders (such as Bougainville in Papua New Guinea and Slovenia), then the drive for ethnic purity will create fewer disturbances than when ethnic groups are mixed (such as in Moldova or Georgia) or if the populations reside in a mosaic pattern (such as in Bosnia-Herzegovina).

Political Components

Collective Direction

Often cohesion within societies is provided in part by a sense of collective direction, which includes the sharing of future goals and the method to achieve them. Such a clear sense of purpose is difficult to achieve and even more difficult to sustain. There are, however, numerous examples in modern history of this cohesion. First, during the independence struggles against imperialist powers, populations and leaders across Africa and Asia did not doubt the desirability of self-rule. Second, during the breakup of the USSR, there was no doubt as to the desirability of multiparty democracy in countries that had been bound by decades of communism. Third, there is little opposition presently in Central and Eastern Europe with respect to the desire to "join" Western Europe and to define their position within all entities associated with Europe (the European Community [EC], NATO, et cetera). These three examples tended to be viewed as pareto optimal and positive in nature, as the benefits they bring exceed the costs. The same cannot be said for the sense of cohesion and collective direction that exists during nationalist bankruptcy. Such a cohesion is based on ethnicity and the universal exclusion of all that are deemed to be "others." The unity of spirit among leaders and population is forged on the basis of ethnicity and sustained by instilling hostility toward members of other ethnic groups. Such a sense of cohesion is based on the identifying of the enemy and the rallying of the population behind the call to destroy the enemy or be destroyed.

However, when this sense of collective direction is ethnic in orientation and largely consumes the energies of the society, it has various dire consequences. One of these is that it becomes a threat to democracy. Indeed, democracy is based on equality among peoples, which is contradicted by social views that one ethnic group has greater rights on a territory than another group. Thus in two of the three examples of societal cohesion that were given above, national consensus disintegrated once the stated goal was achieved. States that achieved independence from colonial powers soon thereafter succumbed to nationalist pressures for interethnic divisions (namely India, Nigeria and the Congo). Furthermore, in numerous postcommunist states, hopes for democracy were accompanied with high expectations that subsequently gave rise to large-scale disappointments. In fact, in both Slovenia and Lithuania, the postcommunist governments were voted out of office after they achieved their independence goals. Such a disintegration of societal cohesion may portend what Gati called the "Weimar syndrome," in which people "will turn from freedom altogether and embrace authoritarian rule, transforming their disillusionment with performance into a rejection of democracy itself."[37]

Law and Order

Nationalist bankruptcy entails the breakdown of law and order within a society as well as the inability of the central government to control it, as is presently the situation in Somalia, Angola and Bosnia-Herzegovina. The source of this breakdown is found at the levels of both the government and the individual. In the case of the former, there exists the widespread application of the concept of "the end justifies the means." If the end goal of nationalist governments is to elevate the status of one ethnic group relative to others, the means to achieve this may include semilegal and criminal activities. Outright government involvement may not be as common as lending support to organizations to indirectly achieve such goals. Such activities are evident in Iran and Libya, and in a different way, in Somalia and Angola. To achieve goals based on ethnicity, governments may support activities such as the forceful eviction of people on the basis of ethnicity or rape and destruction in order to pressure undesirable ethnic groups to relocate. They may tolerate pillage as reward. They may support terrorist activities outside of their state.

During nationalist bankruptcy, individual crime also rises. Some of it comes about because of ethnically based government policies that result in large-scale population migrations and the creation of refugees. These refugees become the objects of resentment of the local population (and thus the victims of crime), while at the same time they are forced to resort to semi-illegal activities themselves in order to assure survival (and thus become the perpetrators of crime). Moreover, the dire economic conditions associated with nationalist bankruptcy aggravate criminal activity as they lower the threshold at which otherwise honest individuals will engage in illegal activities. The extreme becomes acceptable, as theft, appropriations and sanction busting become justified by the existing conditions.

External Environment

Marginalization

Nationalist bankruptcy thrives in an environment of international marginalization insofar as such an environment reinforces xenophobic tendencies of the nationalist governments. This marginalization of a state or region may manifest itself in terms of economics, politics or geo-strategy. In other words, due to changes in the international environment, a region may cease to be of economic, political or strategic importance. The end of the cold war has resulted in the marginalization of numerous regions of the world. The Third World in general has become less important to the West, Afghanistan has lost its role in the cold war theater, Yugoslavia has outlived its role as the maverick communist state, Greece has lost its importance to the EC as the western outpost in the Balkans, and so on. All of these states have had to adjust to their new lack of international importance, which often translates into the drying up of funds that previously poured in. Thus marginalization results in the aggravation of the economic conditions that already are a part of nationalist bankruptcy. Moreover, the international cold shoulder stimulates further xenophobic sentiment among the leading ethnic group, as it raises suspicions that "other" ethnic groups might benefit from the new order.

Intervention

Nationalist bankruptcy thrives in societies where some form of outside intervention exists, be it overt or covert. Usually, intra-ethnic conflicts and animosities within states are considered a "family matter" by the international community, leading to great hesitancy in foreign involvement. However, sometimes foreign involvement does take place, in a variety of forms. It may be bilateral or multilateral; in the latter case it often involves the United Nations (as in the failed secession attempt of Katanga and currently in former Yugoslavia) or more recently the EC (as in Yugoslavia). It may be motivated by irredentism (such as Pakistani efforts in Kashmir or the threatened Turkish efforts in Azerbaijan)[38], or it may be based on political interests (such as Iranian support of Iraqi Kurds) or historical ties (such as German efforts on behalf of Croatia). International intervention may be limited to the sending of food and nonmilitary supplies (such as to the Sudan) or military (such as the Indian intervention on behalf of Bangladeshi secession from Pakistan).[39] Intervention is sometimes limited to passive lack of support of secessionists (such as the policy of the United States during the attempted secession of Biafra) or the withholding of international recognition (such as the procrastination of the EC in recognizing Macedonia). International intervention is sometimes symmetrical, such as when two adjoining states support each other's ethnic/secessionist movements. The most glaring example of such a pattern is that of Sudan and Ethiopia: the Khartum government supported the rebels of Eritrea and Tigre by allowing them freedom of operation on their territory while Ethiopia provided training and supplies to the Sudanese People's Liberation Army. This situation reflected a virtual indirect war, or war by proxy, between the two states.[40]

Regardless of its source, intervention tends to aggravate nationalist bankruptcy insofar as it either lends support for one ethnic group in its efforts against another (thus by implication condoning nationalist government policies vis-à-vis other ethnic groups) or it justifies antagonistic sentiment by the ethnic group that perceives itself harmed by the intervention.

Taken singly, none of the above components would provide the necessary or sufficient conditions for nationalist bankruptcy to occur. Nor are they concomitants that of necessity occur in union. Neither nationalism alone nor economic decline alone cause nationalist bankruptcy. Indeed, a society whose economy is simply failing is not experiencing nationalist bankruptcy. Witness Israel, Bolivia or Germany when they were beset by galloping inflations or Greece when its debt overcame its economy or China when the cultural revolution paralyzed its economy. It is when a large number of the above components are present that they indicate the intractability of the pursuit of nationalist policies and reveal the bankruptcy of goals and aspirations of the leading ethnic group. Various forms of nationalist bankruptcy are today engulfing countries such as Sudan, Somalia, Afghanistan and former Yugoslavia. These regions have all undergone economic decline and interethnic conflict. They have clearly become marginalized with respect to the principal cold war actors, and some have new patrons in the international community that foster and aggravate the situation by taking sides in civil conflicts. This book will focus on the manifestation of nationalist bankruptcy in the former Yugoslavia and its neighboring Balkan states.

ECONOMIC AND SOCIAL CHARACTERISTICS OF THE BALKANS

The mere word *Balkan* evokes an immediate response. In political jargon, it has become synonymous with disorder. *Balkanization* connotes disintegration. The Italian phrase for fruit salad, *macedonia di frutta,* is translated as "fruit macedonia," an allusion to the ethnic mixture of peoples residing in this region of the Balkans. Further, the Serbo-Croatian term for stew, *bosanksi lonac,* translates into "Bosnian pot," in reference to its numerous ingredients. The term *Balkan* is the Turkish word for "mountain," which hints that geography isolates villages and produces the secrecy and distrust that are part of the stereotypic Balkan character. Indeed, the word has come to have such negative connotations that authors have recently apologized to countries for the offense of including them in the Balkans.[41]

Although the connotation of the term Balkan is clear, the geographical definition of the Balkans is not agreed upon by all users of the term. While Metternich is said to have claimed that the Balkans begin at the Rennweg (the road leading south from Vienna), others have focused on regions further to the south.[42] Although some countries clearly belong in this category, the inclusion of others is questionable: Albania, former Yugoslavia and Bulgaria constitute the core of the Balkans. Greece and Romania are usually included. Turkey's inclusion is less certain, since it spreads into Asia: Its western part is geographically and historically a part of the Balkans, especially due to the

Map 1.1
The Balkan States, 1989

ethnic continuity of the Turkish population in adjoining Bulgaria. However, its eastern part is clearly a part of the Muslim crescent that extends into Asia and the Muslim former Soviet republics. Some have argued for the inclusion of Hungary, since it has so deeply influenced the history of the region. Although scholars have disagreed over the inclusion and exclusion of various states, for the purposes of this study, the following are included: former Yugoslav republics, Romania, Bulgaria, Greece and Albania. According to Sjoberg and Wyzan, "Despite the great diversity of [Balkan] experiences, however, one should not lose sight of the fact that the Balkans are once again an integral unit whose component states have sufficient in common to make the concept meaningful."[43]

It has been argued that despite their shared geographical space, the Balkan states have had such different distant and recent histories that it is illogical to lump them together. Indeed, the southern half of the peninsula was under Byzantine and Ottoman influences, while the northern regions were under Austrian and Hungarian rule. Later, between World War II and 1989, the differences among the Balkan states were significant, as each possessed a unique characteristic that set it apart from its neighbors and tested the concept of the Balkans as a unit. Yugoslavia was the maverick of the Eastern Bloc, experimenting with workers' self-management and regional decentralization that left a minimal role for the central government in the economy. With respect to foreign relations, Yugoslavia's break with the Soviet Union, followed by its pivotal role in the nonaligned movement, lent it a special status that Tito aptly manipulated for economic and political concessions. Romania distinguished itself by two seemingly contradictory characteristics. It pursued a foreign policy independent of Moscow, fostering diplomatic relations and pursuing political ties that were the envy of other Warsaw Pact members. Concurrently, its domestic policy was repressive, both politically and economically. In his suicidal and compulsive pursuit of the obliteration of its foreign debt, Ceausescu imposed austerity measures more severe than any conceived by the IMF. Neighboring Bulgaria persisted in its role as the single most loyal political and economic ally of the Soviet Union, the only Balkan state to follow its centrally planned model meticulously, with minor and unimpressive attempts at reform. Albania, the last communist neighbor of Yugoslavia, has consistently created controversy in excess of its size. It remained a Stalinist bastion until 1989, isolated from the world politically and economically except for its monogamous links to first the Soviet Union and then China. Greece was the only noncommunist Balkan state, relying on capitalist production methods and a Western-style democratic political system. It was a member of both NATO and the EC, a fact due more to its geographical location than its economic and cultural affinity with Western Europe. Among these Balkan states there were also differences in the treatment of minorities, an issue especially relevant in 1994. The range included the highly liberal and tolerant attitude of the Yugoslav government, which condoned cultural and political rights of minorities, as well as the highly repressive methods used in Bulgaria in their efforts to Bulgarize Turks and the involuntary dispersal of the Orthodox population of Albania.

Despite these differences, the Balkan states have since 1989 shared numerous similar experiences that warrant studying them together. They are all experiencing an upsurge in nationalism at both the state and substate levels, they are all undergoing economic decline, they are struggling with institutional changes in their economies, they are tied by apprehension of the Yugoslav wars and the historical ties that may be revived. This commonality, together with the geography that determines to some degree the regional response to international stimuli, justify the inclusion of these regions into one umbrella term, the Balkans.

Table 1.1 contains economic and social descriptions of the region. In the table, Romania, Greece, Bulgaria and Albania are treated as states, while former Yugoslavia is divided into the following regions: Slovenia, Croatia (including the "Serbian Republic of Krajina"), Bosnia-Herzegovina (based on Tito's borders, and therefore including the "Croat Republic of Herceg-Bosna" as well as the "Serbian Republic"), and the new Yugoslavia (which includes Montenegro and Serbia with its formerly autonomous regions, Kosovo and Vojvodina). Some of these delineations warrant elaboration and justification. First, it was difficult to decide whether to treat the Serbian Republic of Krajina, the Croat Republic of Herceg-Bosna and the Serbian Republic (of Bosnia) as separate regions. These are de facto states, with individual ruling structures and functioning administrations, yet they are not internationally recognized. The Titoist Bosnia-Herzegovina was recognized by the international community in the spring of 1992, and by the fall of the same year, it ceased to exist as a state: at the time of writing, some 90 percent of its territory is held by Bosnian Serbs or Bosnian Croats that have declared independent regions, and the presidency, led by Alia Izetbegovic, has limited control and power over a small number of people and an even smaller number of institutions. Even the region previously under his control is fractured, as the Bihac region seceded in September 1993 and formed the Autonomous Region of Western Bosnia. Should it thus be treated as a state in this analysis? Despite clear indication that the short term future of Bosnia contains a division according to ethnicity, until such a time, and until the borders of the three ethnic republics are clearly delineated, Bosnia-Herzegovina will be treated as a unit. Second, the new Yugoslavia is defined as Serbia and Montenegro, despite internal pressures to drop that name due to its former connotations and despite its lack of international recognition.[44] Furthermore, at the time of writing, Montenegro is still a part of the union, despite the cyclical demands that it too leave the federation.[45] Third, there is a question as to the justification of including the formerly Hapsburg regions of Yugoslavia in this analysis. Slovenia and Croatia have strong historical ties to Austria, which they have recently accentuated in an effort to set themselves apart from the Balkan states. Nevertheless, Croatia is included in this study

Table 1.1
Characteristics of the Balkans

region	territory (1000 sq. km)	population (million)	GNP per capita (US$, 1990)	literacy (%, 1985)	activity rates (%)
Albania	29 (a)	3.3 (a)	1,200 (b)	na	48.3 (d)
Bulgaria	111 (a)	8.8 (a)	2,250 (a)	93 (c)	49.8 (d)
Greece	132 (a)	10.1 (a)	5,990 (a)	93 (c)	38.2 (d)
Romania	238 (a)	23.3 (a)	1,640 (a)	96 (c)	46.3 (d)
ex-Yugoslavia	256 (a)	23.8 (a)	3,060 (a) 2,480 (f)	92 (c)	45.5 (d)
Croatia	57 (e)	4.9 (e)	3,230 (f)	94.4 (e)	45.6 (e)
Slovenia	20 (e)	1.9 (e)	5,918 (f)	99.2 (e)	50.3 (e)
Bosnia-Herzegovina	51 (e)	4.5 (e)	1,573 (f)	85.5 (e)	38.7 (e)
Macedonia	26 (e)	2.1 (e)	1,499 (f)	89.1 (e)	41.8 (e)
Yugoslavia (Serbia & Montenegro)	102 (e)	10.5 (e)	2,238 (f)	90.0 (e)	S: 45.4 M: 34.4 (e)

Labor force as a percentage of total population.

(a) The World Bank, *World Bank Development Report 1992,* New York: Oxford University Press, 1992, table 1, pp. 218-19. Refers to 1990.

(b) United Nations Development Programme, *Human Development Report 1991,* New York: Oxford University Press, 1991, table 2, p. 123.

(c) Ibid., table 3, p. 124.

(d) Ibid., table 35, p. 183, refers to 1988-89.

(e) Various Tables in Savezni Zavod Za Statistiku, *Statisticki Godisnjak Jugoslavije,* Belgrade, 1990. Population statistics refer to 1989, while activity rates and literacy rates to 1981.

(f) These data refer to 1988, and are presented in Wei Ding, "Yugoslavia: Costs and Benefits of Union and Interdependence of Regional Economies," *Comparative Economic Studies* 33, no. 4, Winter 1991.

because of its link with the Serbian lands to the east by way of the Serbian-held regions of Krajina. As long as Croatia claims this land, and the resident population remains largely Serbian, Croatia will remain tied to the Balkans. In addition, its economy is tightly interwoven with that of the rest of Yugoslavia, further bonding it to the Balkans. The inclusion of Slovenia is more questionable, as its severance from the Balkans can be achieved cleanly. Nevertheless, it was included in this study because of its former link to Yugoslavia.

With respect to territory, the former Yugoslavia was the largest country of the Balkans (see table 1.1). In the aftermath of its dissolution, Romania took the lead, followed by Greece, Bulgaria and the new Yugoslavia. With respect to population, Romania and the former Yugoslavia were also close, while since 1991 Romania has been the most populated Balkan state. However, with respect to GNP per capita, Romania's position drops and that of Greece is elevated, followed by Slovenia. Indeed, Romania's state income is even lower than Bulgaria's and not much higher than Albania's. Literacy rates everywhere are relatively high, although they mask the wide variation between the rural and urban regions. Activity rates are similar across the Balkans, although there are differences between male and female employment rates, especially in Albania and Greece. Romania, Bulgaria and the more developed former Yugoslav regions consistently experienced higher female activity rates.

Nationalist Bankruptcy in the Balkans

While there is no doubt that nationalism as a strong force has emerged everywhere in the Balkans (see chapter 2) and that there has been severe economic decline (see chapter 3), the evidence of nationalist bankruptcy is strongest in the former Yugoslavia. There, the economic crisis of the 1980s produced interethnic turmoil that thrived in an accepting international atmosphere. Indeed, Dyker notices in the former Yugoslavia: "We have seen large sections of the Yugoslav population . . . turning their faces away from the real problems of the day, retreating into an atavistic ethnic sectarianism which can only create new barriers to the national consolidation which Yugoslavia so desperately needs."[46]

Yugoslavia may be viewed as the epicenter of an earthquake that is shaking the entire region, and thus, whether the war spreads to adjoining non-Yugoslav regions or not, elements of nationalist bankruptcy are visible at this early stage across the region. Greece seems most removed from this quake insofar as its economic system is least affected. However, its nationalist concerns seem to be rising as the international situation is changing, and its insecurity vis-à-vis the rising power of Turkey has permeated the nation. According to Perry, "In Greece, where economic decline has rendered people generally more receptive to intensified chauvinistic rhetoric, nationalists [emerged]."[47] Moreover, events in Greece in August 1992 led to the following report: "[Greece has the] image of a country at war, in which daily events (such as turning on the light, driving a car, mailing a letter and cashing a check) have, for weeks, become a laborious task."[48] This is because Greece has entered in 1990s with a severely weakened economy: at the end of the 1980s, Greece was facing 15 percent inflation, public sector borrowing at over 20 percent of GDP, and a large current account deficit. The introduction of policies by the conservative government that came to power in April 1990 has caused further havoc in the economy and among the population.

Romania and Bulgaria both face economic difficulties associated with the demise of communism. In both states, nationalist governments that are appealing to the ethnicity of the majority have taken power, while multiparty politics has brought forth nationalist political parties through which minority ethnic groups attempt to protect their rights. These groups include primarily the Hungarians in Romania and the Turks in Bulgaria. Albania is perhaps in the worst economic straits, described by *The Economist* as the "basket case of Europe."[49] According to Glenny, "The Albanian revolution was accompanied by the rapid atrophy of the social infrastructure countrywide, such that living standards had dropped below the level at which normal social psychology begins to function."[50] Nationalism has reached its peak in the affinity that the Albanians feel towards their brethren in Kosovo.

The characteristics of nationalist bankruptcy enumerated in the previous section are described for the Balkan states:

Economic Decline

That there is an economic crisis in the Balkans in the early 1990s is neither surprising nor new. Indeed, to speak of an economic crisis today in this part of the world is almost trite, as Yugoslavia has been said to be in perpetual crisis since the 1980s. Greece has been expecting an economic improvement for so long that its development record is called "the takeoff that never was." Albania was predicted to have "gone under" numerous times in its postwar history, while many Romanians would actually claim that they have indeed gone under. Why is the present crisis any different from those experienced in the previous decades? It is, in fact, deeper and broader. GDP dropped precipitously during 1992 (Albania, -21.1 percent; Bulgaria, -25.5 percent; Croatia, -30 percent; Greece, -0.8 percent), while unemployment soared (Romania, 9 percent; the new Yugoslavia, 19 percent; Greece, 8 percent) and inflation flared (Bulgaria, 100 percent; Romania, 250 percent; Greece, 18 percent; Macedonia, 70 percent monthly; Bosnia-Herzegovina, 86 percent monthly). Moreover, economic decline is noted in other macroeconomic indicators (such as indebtedness and the bulging underground economy) as well as microeconomic ones (such as insufficient capital accumulation, failing consumer goods markets and price distortions). All this has led Cviic to call the Balkans "an economic black hole."[51]

While the particular configurations of the economic decline may vary across the Balkan states, some sources of the crises are shared, including (i) the Yugoslav war, (ii) the transition from socialism to capitalism and (iii) economic trends in the international economy. With respect to the Yugoslav war, effects are obviously the greatest in the regions directly involved in the fighting, as their economies have become transformed by the demands of war: the labor force is disrupted, production has turned toward war goods and the state budget is strained to pay for war-related activities. All former Yugoslav regions have suffered from the influx of refugees and the loss of human capital associated with the out-migration of skilled labor. The new Yugoslavia is crippled by sanctions, which have produced an abundance of misery and lawlessness. The neighboring Balkan states are also affected by the sanctions, insofar as traditional trade routes have been disrupted, and many have had to demand compensation from the UN and the EC.

With respect to the transition process, the formerly communist states are struggling, with varying degrees of commitment and success, with privatization, market prices and institutional change. Greece, the only noncommunist state, is also undergoing a transformation of its economy, as the new austerity measures introduced under the IMF prescription (mostly aimed at trimming the public sector) have proved extremely unpopular and have shaken the social and economic systems. With respect to international economic trends, there is no doubt that the Balkans have become increasingly marginal in global affairs, that the expected western investment failed to materialize and that the domestic pressures faced by Germany, the EC nations and the United States are such that few resources are left to satisfy the massive demands of the Balkan societies and economies. Furthermore, the dissolution of the Committee of Mutual Economic Assistance (CMEA) and the loss of trading partners has had a dramatic effect on the Balkan states, especially on Romania and Bulgaria. Also important for their economies was the 1990 rulings that CMEA trade take place in foreign currency and that tax credits are no longer acceptable. The economies of all Balkan states are also affected by the general recession in Western Europe and the United States, affecting the demand for exports and hindering foreign investment. Lastly, the Balkan economies are also adversely affected by the Gulf War, for two reasons: first, the resultant higher price of oil implied that import expenses increased (at least in the short run) and second, Iraq has failed to pay up on its credits and jobs completed. This latter is especially true with respect to former Yugoslavia, which had cooperated extensively with Iraq on engineering projects for which the bills are still outstanding.

Economic Decline and Nationalism

Nationalism may cause leaders and populations to make decisions and take steps that are not in the best interest of their economies. Thus instead of improving the economic picture, nationalist policies may in fact aggravate it. Evident all across the Balkans, examples of this abound in the former Yugoslavia. Numerous scholars have shown the harmful effect on the Yugoslav economy of nationalism in decision making. Ramet's study of Yugoslav nationalism indicates the effect of nationalist pull on decisions pertaining to the building of ports and railways and the expansion of airlines.[52] Glenny notes that nationalism had an effect on the construction and upkeep of the highway connecting Belgrade and Zagreb.[53] Cochrane discusses the decentralization according to region associated with an ethnic group and the severe effect on agricultural production of the waste associated with duplication of services.[54] Djilas predicted that such characteristics of nationalism would in fact be the source of the demise of Yugoslavia: "This [Yugoslav nationalism] is not classical nationalism, but a more dangerous, bureaucratic nationalism built on economic self-interest. This is how the Yugoslav system will begin to collapse."[55]

With respect to other Balkan states, there is also significant evidence of nationalism in economic decision making. Greece is imperiling its standing with the EC over the Macedonia issue; Albania is trying to improve its standing with Muslim countries by supporting demands for an independent Kosovo while rupturing links with its neighbors and immediate economic partners; Bulgaria pursued policies of expulsion of Turks despite the negative effect that would have on its relations with Turkey and the West; Romania followed nationalist policies in Transylvania by encouraging industrialization in an effort to appease the Hungarians while at the same time diluting the local population by a large influx of ethnic Romanians.

The rise to power of nationalist leaders in Croatia, Serbia and the adjoining states is evident, although their futures are unclear if they fail to deliver on the economy. The ruling administration in Serbia has a ready-made scapegoat, namely the sanctions, to exonerate them of economic mishandling. Croatia also has the war with the Serbs in Krajina as an excuse for the disruption of the foreign currency inflow from tourism. However, the adjoining Balkan states lack such culprits. There, widespread nationalism may be shortlived if it does not deliver on the economy. Lithuania is a case in point, as in October 1992 communists were voted back into power because the nationalist Sajudis party simply could not improve the economy. The DEMOS coalition in Slovenia suffered a similar fate when it was unable to deliver on the economy.

Nationalism and Territorial Goals

The fact that nationalism has so prevailed in the postcommunist world, including in the Balkans, gives rise to this question: Is it that the communist system did not give legitimate expression to its minorities' manifestations of nationalism, leading to the present obsessive nationalistic passions, or was there too much liberty to express nationalistic tendencies, as in Yugoslavia, preventing the creation of a sympathetic sentiment toward the center?[56] Many claim that with the demise of communism, nationalist sentiment could be freely vented. Isaiah Berlin claims that "in our modern age, nationalism is not resurgent: it never died. [It and racism] are the most powerful movements in the world today, cutting across many social systems."[57] O'Brien claims that democracy and nationalism are incompatible, calling nationalism a "conglomerate of emotions" and claiming that "the early stirrings of nationalism appeared to be democratic, but later manifestations were more disquieting."[58] Under communism, nationalist feelings were dormant not only within countries, but among countries also: for example, Janos Kadar in Hungary ignored the Hungarians in the diaspora until the mid-1980s, when the importance of the diaspora increased in government rhetoric. Such irredentist sentiment among nationalists was not uncommon in states that had hitherto accepted the post-World War II boundaries as an inherent part of the status quo, part of what Cviic calls the "cold war corset."[59]

Whatever the causes, a form of nationalism is emerging at both the state and substate levels across the Balkan countries. In repressive Albania, in decentralized Yugoslavia, and in the western democratic tradition of Greece, nationalist leaders have gained great appeal among the populations as they spew forth statements reflecting national introversion and underscoring the superiority of one group relative to others. Even at the substate level, there are a plethora of ethnic groups vying for attention with demands varying from secession (as in Slovenia), to increased cultural rights (as in Transylvania), to an adjustment of boundaries (as in Cyprus and Bessarabia). The coming to power of nationalist leaders in the postcommunist era has been all too common. In former Yugoslavia, the ascendancy of nationalist leaders may have been cloaked in different mantles; however, presidents Tudjman,

Milosevic, Kucan, Izetbegovic, Bulatovic and, to a lesser degree, Gligorov all used anticommunism to enhance their legitimacy. Fundamentally, they were all nationalists bolstering the "us versus them" distinction. The ruling parties of Serbia (Socialist Party of Serbia), Montenegro (the Democratic Party of Socialists) and Romania (the Democratic National Salvation Front, renamed in July 1993 the Party of Social Democracy) seem to represent transformations of communist parties into nationalist parties. Romanian president Ion Iliescu is the leader of the Democratic National Salvation Front, a party led by former communists. He and his party were the reason why *The Economist* stated: "Old communist parties never die, they just find new ways of exerting their influence."[60] The nationalist trend was less strong in Bulgaria and Albania, although pressure for it grew with the precipitous events in neighboring Yugoslavia. Indeed, in Albania, the ruling Democratic Party has begun to vilify the opposition by claiming it is "anti-national."[61] Such nationalism has been evident in the ruling circles of Greece also. Indeed, the vehemence with which the government responded to the issue of recognition of Macedonia brought to the forefront Greek nationalism, which had been largely dormant since the Cyprus conflict of 1974. In addition, the creation of the Spring Party as a rebellion to the insufficient nationalism of the Conservative government indicates the trend is getting stronger rather than weaker.

In the Balkans, the nature of the nationalism has been largely territorial, and the quest for the creation of nation-states and adjustment of boundaries is supreme, as discussed in chapter 2. This strong territorial component has led Zaslavsky to differentiate between the concept of nationalism and ethnic mobilization and to claim that what is present in the former communist societies is the latter.[62] In the former Yugoslavia, the regions of Slovenia, Croatia, Bosnia-Herzegovina and Macedonia seceded from Yugoslavia and attempted to retain their administrative, Titoist boundaries. Slovenia is presently experiencing boundary conflicts with both Italy and Croatia. Croatia is dealing with territorial questions with the Krajina Serbs, as well as autonomy issues with populations of Istria and Dubrovnik. Bosnia-Herzegovina is being torn apart by three ethnic groups, only one of which purports to want to retain its multiethnic fabric. Macedonia is facing autonomy and territory claims from the ethnic Albanians who have irredentist aspirations. Serbia is experiencing three territorial pulls: the Hungarians in Vojvodina, the Muslims in Sandzak and the Albanians in Kosovo. Even tiny Montenegro is not immune to territorial pretensions, as the Croats and Albanians dream of the reestablishment of wartime borders of Croatia and Albania; during the war Mussolini and Hitler extended the borders of its allies Croatia and Albania to include parts of present Montenegro. In Romania, two territorial trouble spots exist: Transylvania and Bessarabia. The issue of the future of both of these regions has mobilized a large number of people and political parties. In Greece, the question of recognition of Macedonia and the long-term borders of that state has shaken the country to the core. It has unleashed renewed demands for control over North Epirus (presently part of Albania) and the resolution of the Cyprus crisis. It led to the splitting up of the leading party, the New Democracy Party, from which former Foreign Minister Antonis Samaris left to form the Political Spring Party, whose platform includes a tougher stand on issues concerning Macedonia, Albania and Cyprus. Albania is also concerned with the territorial question of neighboring Kosovo, where it sponsors irredentist activity.[63] At this time, it seems that only Bulgaria does not openly foster territorial claims, nor does it have minorities that are making ethnoterritorial claims. As discussed in chapter 2, the Turks and the Pomaks are not territorially compact, having been the subject of numerous dispersals and dislocations since the Balkan wars.

In their quest for territory, preferably to be ethnically cleaned of "non-desirable" minorities, the nationalist leadership in much of the Balkans has introduced nationalism into the constitutions and thus institutionalized an unequal position for minorities. It is a short-term move that is easier to achieve than territorial aims. Hayden's study of the new constitutions of the former Yugoslav states indicates that some new Balkan constitutions in fact demote ethnic groups. In some Balkan countries, the lack of some minority rights was evident long before the present surge in nationalism: Indeed, the refusal to recognize minorities such as the Macedonian Slavs in both Greece and Bulgaria is a form of constitutional nationalism. The Greeks call them Hellenic Slavs while Bulgarians call them Western Bulgars. Albania's denial of basic rights to ethnic minorities led to the blocking of its entry into the European Democratic Union.[64]

Armed Struggle

From the Western perspective, it seems that the Balkan populations are all too ready to take up armed struggle for ethnic and territorial causes. The speed with which former Yugoslavia blew up and the ferocity with which neighbors are slaughtering each other remains a mystery to the western populations, which seem less moved by questions of ethnicity. Even Yugoslavs wonder at the ferocity. Its source is explained as being "the inter-generational socialization of negative stereotypes regarding the history and behavior of other groups."[65] However, the present war has attracted so many combatants not because of negative stereotypes, but because of the citizens' desire to protect their property rights and their families. Draft evasion in both Serbia and Croatia was evident only when the fighters were conscripts into the army that were being sent to territories other than their own.

Willingness to take up armed struggle is presently also evident in Albania, where the government has consistently made statements indicating commitment and readiness to partake in the "liberation" of neighboring Kosovo. It is evident in the disputed territories of Bessarabia, where the Moldovians, the Dniesters and the Gagaus have readily attacked each other, and the Transcaucuses, where seven wars are presently raging. Greeks and Turks are still facing each other with guns in Cyprus. Indeed, in the course of this century, the Balkans have experienced more armed struggles than any other part of Europe.

Large-Scale Population Movements

The Balkans have throughout history experienced large-scale population movements, both in and out of the region as well as within it. The twentieth century began with the population shifts caused by the Treaty of Versailles, whose terms placed people on the wrong side of borders. Thus Hungarians, Turks and Bulgarians relocated in large numbers. In the early 1920s, some 2 million Turks and Greeks exchanged countries of residence. World War II brought on similar movements of peoples, including the exodus of the Germans, and the relocation of Hungarians and Albanians. Today, some 2 million people from former Yugoslavia have relocated to temporary homes, awaiting the resolution of the conflict and their fate.

Some of these movements have produced greater peace in the region, such as the departure of Germans at a time when their presence would have kindled revenge. Similarly, the relocation of Turks and Greeks, while costly, produced peace in the region. So too, for the sake of long-term peace and security, there is likely to be a large-scale permanent population exchange in former Yugoslavia. Other population movements, such as the exodus of Serbs from Kosovo, the inflow of Albanians into the region and movements of the Turks out and back into Bulgaria in the 1980s, have created great animosity and thus instability among the Serbs and Bulgarians throughout the Balkans.

These large-scale population movements are extremely costly, both for the losing and the receiving regions (as described in chapter 6). Thus their effect on the economies of the Balkan states is large, especially in the former Yugoslavia.

Not only do all the former Yugoslav republics have to deal, to varying degrees, with the upkeep of the refugees, but Serbia and Croatia have to deal with the loss of skilled workers and concomitant brain drain. Such events further accentuate the economic crisis and exacerbate nationalistic feelings.

Collective Direction

While there may have been a popular desire across the formerly communist Balkan states in the late 1980s to "join Europe" and in the process divorce themselves from the communist system, this sentiment seems to have been less strong and less consistent than in northern states such as Poland and Hungary. Zaslavsky calls this bond to Europe and the feeling of belonging to Europe the "new nationalist myth which unites and animates both the leaders and the rank-and-file members of these nationalist movements."[66] In reference to the East European countries, Ash says that their "basic model is the European community—the one and only, real existing common European home."[67]

After the demise of communism and the breakup of the Soviet Union, the path to be followed for the future was not clear. Wavering over future direction occurred with respect to two issues: first, the pace and extent of economic reforms and the subsequent embrace of capitalism, and second, the pace and extent of political reform and the embrace of a western democratic system. With respect to the former, Balkan societies differed among themselves and among their populations in their support for liberalization and marketization. Although people believed that short-run pain had to be endured in order to experience an improvement in the long run, the decrease in the standard of living that the transition to capitalism imposed on them tried their patience. A sense of nostalgia for past security and stability began emerging. Indeed, polls in Romania in 1992 indicate that some 47 percent of the population perceived life to have been better under Ceausescu.[68] Moreover, 43 percent said that the path Romania was taking in 1993 was incorrect.[69] In Bulgaria, 65 percent of the population was dissatisfied with the standard of living in fall 1992.[70] In former Yugoslavia, media interviews clearly show that people are disillusioned with what the demise of communism has brought.[71] Indeed, a study of all former communist states indicates that only in the Czech Republic did the population feel better about their lives and the economy in 1993 than in 1989.[72] This sentiment sets the stage for the emergence of authoritarian leaderships that are populist in nature and unwilling to bring about the painful economic transformations necessary to uplift the economies, producing, in the words of Aslund,

"Latinamericanization, implying populist authoritarian policies with lasting macro-economic imbalances."[73]

The second issue over which there is wavering among Balkan populations and leaders is political liberalization. While there is no doubt that a relatively abrupt opening up of the political debate, reflected in the emergence of numerous and varied political parties, manifested itself across the entire communist Balkan region, after the dust settled, it is unclear how much actual debate and choice is left. In many countries, one or two major parties have monopolized the entire political spectrum. In Serbia, the ruling Socialist Party, under the leadership of Slobodan Milosevic, presently has as its only significant opposition the Radical Party, under the leadership of Vojislav Seselj. In the aftermath of the 1992 elections, the Socialist Party controlled 31.4 percent of the Federal Assembly while the Radicals held some 22.4 percent. This concentration of power in two parties is in sharp contrast to 1991, when Serbia boasted over fifty political parties. Croatia has also become a de facto one-party state, as the ruling Croatian Democratic Union (HDZ) controls over half of the parliamentary seats.[74] The elections of February 1993 indicate that the key issue in domestic affairs is the authoritarian presidency of President Tudjman. However, there is dissension within this leading party, and this dissension is likely to result in a rupture within the HDZ. By June 1993, a veritable political battle was shaping up between the hardcore nationalists, mostly from Herzegovina, and the moderate forces.[75] The liberal core is focused around Stipe Mesic, while the conservative faction formed around Vladimir Seks. Polls suggest that the former is three times more popular than the latter.[76]

Romania, Bulgaria and Albania all contain numerous insignificant parties but are ruled by a single strong one. Thus one single-party system seems to have been replaced by another single-party system. In Romania, the National Salvation Front was expected to split in two, like the Greek and Croatian ruling parties. However, this did not happen. Instead, the party continued to attract different people and has failed to renew itself ideologically since it came to power. Another party has become notorious for its more nationalist platform, the Party of Romanian National Unity. Its chairman, Gheorghe Funar, is based in Cluj, where he banned bilingual signs in Hungarian and Romanian, thus pitting his party against the ruling party on this crucial issue.[77] In Bulgaria, the ruling party, the Union of Democratic Forces, under President Zhelyu Zhelev, is so strong that there have been rampant accusations of a political "recommunization" across Bulgaria.

Irrespective of the number of parties and the activeness of the opposition, the Balkans are characterized by the emergence of nationalist leaders who have come to power by accentuating the needs and demands of the majority, usually the titular nationality.[78] This is true both in the postcommunist states and in Greece. In the former, the ruling parties have elevated the principal ethnic group relative to others in a wave of introversion and elitism that seems to contradict their coexisting desires to "join Europe" and to internationalize. In former Yugoslavia, the civil war has undoubtedly served to strengthen nationalist parties both in Serbia and Croatia. In Serbia, elections in December 1992 gave Milosevic's Socialist Party of Serbia over half of the vote. These elections indicate clearly that the overwhelming majority of voters supported the nationalist policies of either Milosevic or Seselj, while a contender such as Milan Panic had little chance in an atmosphere where there was little inclination for compromise and concession.[79] Milosevic rose to power in response to a popular need, articulated by the now famous 1986 Memorandum of the Serbian Academy of Sciences and Arts, in which Serbian grievances against Titoist Yugoslavia were enumerated and exposed. This led to a wave of breast-beating across the Serbian lands. Indeed, Djilas points out that "the foremost creators of this bitter national ideology were the intellectuals."[80] In Croatia, Tudjman first came to power as Croatia was riding a wave of Croatianism, and he promised all the appropriate accoutrements: independence from the Yugoslav federation, emergence as a central European power and acceptance into the global community of nations, in addition to a great deal of symbolic paraphernalia.[81]

The Slovenian coalition DEMOS, consisting of six parties, brought Milan Kucan to power. Despite the relative ease with which Slovenia both left the Yugoslav union and was accepted into the world community, it was necessary to foster and play up the Slovenian national spirit. This phenomenon has been described: "Many Slovenian commentators have remarked on the ...post-communist emphasis on the 'Slovene essence,' which includes references to 'Slovene chickens,' 'Slovene soil' and 'Slovenian descent' of families. Other observers, however, have pointed to the potential dangers of calls for the banning of all that is multi-cultural and non-Slovene."[82] In Macedonia, similar accentuation on the Macedonian personality, the Macedonian character, the Macedonian church, language and culture permeated the functioning of the society, and the enthusiasm with which this newfound Macedonian-ness was embraced served to alienate not only the non-Macedonian local populations (namely the Albanians) but also the Greeks, as Macedonian effervescence overflowed as far as neighboring Thessalonika.[83] In Bosnia-Herzegovina, the leaders of the three ethnic factions are of necessity nationalists. The Croat Mate Boban and the Serb Radovan Karadzic never hid this fact. The president of Bosnia, Alia

Izetbegovic, claimed to represent a multiethnic society. However, as the war in Bosnia progressed, the effort to represent all sides decreased as ethnic polarization within the former republic grew in magnitude. In Greece, soon after the coming to power of Constantine Mitsotakis, dramatic changes swept its northern neighbors and Greece found itself in a newly precarious position relative to Albania, Macedonia and Turkey. The demands of the new times, coupled with the demands of the population, pushed leaders to take more nationalist positions. Indeed, Mitsotakis has been under criticism both from within the party and from outside. The party has split as Antonis Samaris broke off to form the Political Spring, a party that disagrees with Mitsotakis on the Macedonian issue. The opposition PASOK, led by Andreas Papandreou, has also condemned the ruling party for inadequately protecting Greece's interests and, upon coming to power in the October 1993 elections, promised to make Greek nationalist issues primary.

Thus, across the Balkan states, the sense of purpose and collective redirection has been provided largely by nationalist sentiment, based on the virtual discovery of ethnicity after decades of dormancy and then the elevation in standing of one ethnic group relative to others. Secondary in this atmosphere is the desire to modernize and democratize. Indeed, Cohen remarked that "the most serious threat to the institutionalization of democratic rule . . . has been the salience of nationalism and ultra-nationalism in Balkan political life."[84]

Law and Order

The most severe case of breakdown of law and order has been in Bosnia-Herzegovina, followed by some other former Yugoslav republics. Clearly, Bosnia has suffered from village-to-village fighting in which five different military formations have partaken. The proliferation of military organizations in which the channels of communication and direction are tenuous at best enabled atrocities to take place, including personal damage such as rape, impaling, castration and execution as well as property damage such as pillage, arson and thievery. Serbia and Croatia have not been immune to a changing social climate in which, despite characteristics reminiscent of police states, the centers have not managed to control petty or major crime. Robbery and gang warfare have increased dramatically since 1991, due to the proliferation of arms among the population, the decrease in standards of living and the new opportunities for aggrandizement from illegal activities. Sanctions have given rise to illegal activities in Serbia and adjoining Balkan states, as the benefits from sanction busting outweigh the costs. Minic, who recently described the black-market economy in Serbia, notes that "there has been a progressive infiltration of black-market behavior and a criminal mentality into all reaches of the economy and the society."[85]

Marginalization

From a global perspective, the Balkan states as a whole have become marginalized. All Balkan states had a raison d'être during the cold war order. Yugoslavia was the maverick communist state, Romania diverged from the Soviet Union in foreign policy, Albania was the enigmatic Chinese puppet state, Croatia was a highly Catholic region despite communism and Greece was the only noncommunist state in the Balkans. At the present time, the West no longer needs the Yugoslavia that it created as the maverick within communist states. So too with all other former communist states. Even Greece has also lost its usefulness to both NATO and the EC. This marginalization of the region is clear also in the unwillingness of the West to intervene with serious efforts to end the war in Bosnia. No vital interests of any major power are being threatened, and the likelihood of the civil war spreading into regions where there are vital interests is quite low; therefore serious attention is unlikely. Unlike Kuwait, where the economic lifeline of the West was being threatened, Bosnia offers little compensation and little reward. Perhaps the least marginalized states now are Romania and Bulgaria, as both the Greeks and the Turks are vying for influence there.

Foreign Intervention

Numerous times in the history of the Balkans, big power interference in Balkan affairs was more overt than it is in the present. In the least, it consisted of the extension of tacit support to Balkan states, and at most, it included armed aggression against a Balkan region. Indeed, historically France and Russia are known to have supported Serbia; Austria and Germany have felt an affinity with Croatia; Italy has had a big-brother relationship with Albania; Russia has looked after Bulgaria's interests, and so forth. In the postcommunist Balkans, some of these historical liaisons are reemerging. The most overt of these has been the extension of covert and overt support to Croatia by Germany and Austria. Undoubtedly, Germany's actions in the Yugoslav crisis have contributed more than any single outside factor to the onset of the current bloodshed.[86] It was Germany's support of Croatia and Slovenia *prior* to their recognition in December 1992 that gave rise to the war insofar as it emboldened those states to secede without adequate provision for the Serb population.

Similar support is extended by Turkey to the Muslims across the Balkans, whether of Slavic descent or not. The historical animosity of the Turks toward the Greeks, as well as the Hungarians toward the Serbs, has re-emerged. Russian support of the Serbs is resurfacing, although not to the extent that the Serbs would want. The United States has emerged as an active player in the Balkans by singlehandedly pushing for the imposition of the sanctions against Serbia. Western Europe, on the other hand, is pushing for the retention of the ban against arming the combatants, thus preventing Muslim arms in the region. These forms of intervention, as well as others, are discussed in chapter 7.

Why is nationalist bankruptcy appearing in some Balkan regions in the 1990s? Why are ethnic groups that had until recently cohabited relatively peacefully suddenly beginning to feel that cohabitation is no longer possible? The proximate source is fear. It is fear for one's property and family, for one's ancestral graves and one's history, that leads people to "cleanse or be cleansed." This fear is justified; it is rooted in history, and it has had a basis in events that have occurred in recent and distant history. These events are intricately tied to the psyche and the national fiber of each of the ethnic groups. Aleksa Djilas describes this fear in the case of the Bosnian Serbs: "Bosnian Serbs are undoubtedly greedy to keep as much territory as possible. But this is not the main reason for their obstinacy. The fear of living with Muslims and Croats in any form of a common state is a much more important reason."[87] In addition to the fear factor, nationalist bankruptcy is the result of a combination of various factors: the severe decline in living standards (with both internal and external causes), the emergence of nationalist leaders with xenophobic goals that give vent and scope to the individuals fears, and an international environment accommodating to turmoil because of the lack of principles and knowledge as to how to deal with change. Thus the particular mix of internal and external conditions resulted in an economic and social environment conducive to the rise of nationalist bankruptcy.

The present situation in the Balkans is similar to others in its past. Indeed, aspects of nationalist bankruptcy have manifested themselves previously—the economies dwindled, foreign powers meddled and local ethnic groups bickered. Interethnic, interreligious and intercultural animosities have persisted through history along this fault line. Border disputes have raged throughout history, as ethnic groups have tried to assert their claim to identical territory. Wars, accompanied by atrocities, have been fought to solve recurrent interethnic issues. Indeed, Nobel Prize winner Ivo Andric wrote about atrocities committed by the Turkish rulers in Bosnia, Robert Kaplan wrote about Serbs committing atrocities against the Muslims, the

New York Times in 1917 reported abuse of Serbs in Bosnia: "Compelled to dig their own graves, drowned, burned alive, hanged or shot down with machine guns, the Serbians of Herzegovina, Bosnia, Istria and Dalmatia were victims of Austro-Hungarian atrocities surpassing the human imagination."[88] Thus, Pfaff is wrong to say that the basis for present hatreds lies in the present: "[Yugoslavia's] supposedly primordial hatreds are a twentieth-century phenomenon."[89] It seems instead that in the Balkans, periods of inter-ethnic peace are the exception rather than the rule. In fact, the communist period stands in sharp contrast to other times in the history. Markovic has claimed that it is periods of instability that are the constant in the region, and that the people of the Balkans live in "extraordinary" situations more often than in the "ordinary."[90]

2

Ethnoterritorial Disputes and Self-Determination

"No matter what territorial redistributions might take place, some minorities would be left behind in the reshuffling."

Joseph Roucek[1]

"Nationalities have turned into political parties."

Grigorij Pomerants[2]

"A nation is a people united by a common dislike of its neighbors and by a common mistake about its origin."

George Brock[3]

"If nature abhors a vacuum, politics does not tolerate a no-man's land."

Uri Ra'anan[4]

"If 'democracy' is conceived of as requiring a polity of equal citizens, or at least as requiring the protection of minorities, (ethno)national self-determination is an anti-democratic principle."

Robert Hayden[5]

In preparation for the Paris Peace Conference at the end of World War I, President Woodrow Wilson commissioned Walter Lippmann to lead a group of four in an effort to redraw the boundaries of Eastern Europe. The result of this project, which involved poring over maps and statistics pertaining to the area, was the basis of Wilson's 14-point program. However, many of Lippmann's suggestions, mostly having to do with territorial arrangements, were not heeded, leading him not only to pull out of the project but also to remark: "Looked at from above, below, and from every side I can't see anything in this treaty but endless trouble for Europe, and I'm exceedingly doubtful in my own mind as to whether we can afford to guarantee so impossible a peace."[6] He felt that drawing boundaries that placed large numbers of people in a foreign state was bound to imply future trouble. This view was supported by Guglielmo Ferraro, who in 1925 wrote that the borders of the Balkan states could not survive in the long run.[7] Today, almost 75 years later, the world is once again under pressure to redraw Balkan boundaries, and Lippman's and Ferraro's words might be applied to the various proposed peace plans for Bosnia.

The focus of this chapter is the role played by ethnicity (and nationalism) in Balkan territorial disputes. Indeed, both nationalism and pressures for boundary adjustments are an integral aspect of the national bankruptcy that permeates the region. Ethnically based boundary issues, although always in existence, have achieved a new notoriety in the recent past, mostly due to the breakup of the Soviet Union and its cartographic ramifications. It is estimated that there are presently 76 potential border disputes in the former Soviet Union, most of them based on ethnic claims, and of the 23 interrepublican borders, only three are not contested.[8] In addition, there are presently 32 civil wars raging across the world, many of them ethnically based and territorially motivated.[9] While the past few years have witnessed an increase in ethnically motivated boundary disputes, this does not imply that the creation of new states must be accompanied by ethnic turmoil. Indeed, numerous newly emerging states never experienced ethnic pressure over boundary issues. In fact, regions such as those in Soviet Central Asia never experienced a separatist movement of any kind: Zaslavsky points out that "their independence was not achieved through self-determination but rather imposed on them from outside."[10]

ETHNICITY AND BORDERS

Definitions and Concepts

An ethnic group, according to Narroll, is defined as a biologically self-perpetuating group that shares fundamental cultural values and differentiates itself from other groups.[11] These cultural values may be embodied in language, religion or myth of origin. Which of these predominates is pointed out by Uri Ra'anan: in 'eastern regions,' ancestral language seems to be the dividing factor, while in 'southern regions,' religion is the primary differentiating aspect of ethnicity.[12] Barth has focused his definition of an ethnic group on the boundary that defines it rather than its cultural components.[13] This leads to the connection between ethnic group and territory.

The ethnoterritorial concept relates the ethnic group to a territory. According to Thompson and Rudolph, "the term ethnoterritorial is used . . . as an overarching concept for various political movements and conflicts that are derived from a group of people . . . having some identifiable geographic base within the boundaries of an existing political system. . . . The people must identify themselves or be identifiable as a group distinct in such characteristics as their culture, language, history, religion, traditions and/or political past."[14] Other concepts are related to this. According to Connor and later Shiels, ethnically based separatism and the desire to redefine borders grows out of ethnonationalism. The definition of ethnonationalism is "the sentiment of an ethnic minority in a state or living across state boundaries that propels the group to unify and identify itself as having the capacity for self-government," while ethnic separatism is "the movement by members of an ethnic group to gain autonomy over their own destiny, with the formation of a separate state as the major option."[15] Most secessionist efforts in the world in the 1990s are cases in which ethnonationalism has taken on elements of ethnic separatism. Most of these involve territory. Indeed, the nature of the nationalism in numerous parts of the world transcends the "we/they" distinction based on ethnicity, religion and history, and is highly political with strong ethnoterritorial pretensions.[16] Indeed, as claimed by Pomerants, "nationalities have turned into political parties" that are venting their national aspirations with respect to border changes.[17] In the absence of an ethnic group, pressures for border changes become simple border disputes, of which examples abound in South America, or the result of problems of federalism, such as the southern confederacy that led to the U.S. Civil War. Today, Lombardy represents the most glaring example of a nonethnic attempt at redefining boundaries: northern Italians support a divorce from central and southern Italy, from whose population they do not differ ethnically.

This effort to redefine borders is in part due to the fact that political borders across the globe do not necessarily coincide with ethnic boundaries.[18] A few definitions are in order to illustrate this point: A state is a legal-territorial concept referring to that set of structures and institutions that seek to maintain control over a population within a specific geographical area. A nation refers to a group of people who share culture, history and usually language in a specific territory and who give political expression to this common identity.

(In the English language, "nation" and "state" are used synonymously, while outside of America, "nation" and "ethnic group" are interchangeable). A nation-state is a state in which national and political borders coincide. In most cases, they do not. Indeed, according to Connor,[19] of a total of 132 contemporary states, only 12 are ethnically homogeneous. In 25 states, one ethnic group accounts for more than 90 percent of the population, while in an additional 25 states it accounts for between 75 and 90 percent. In 31 states, the dominant ethnic group represents 50 to 75 percent of the population, and in 39 states, it represents less than half. It is also noted that in 53 states, the population is divided into more than five significant groups. In many of these cases, the multiethnic states have worked out an elaborate system of mutual tolerance and a modus vivendi has emerged among ethnic groups sharing a common political space. This explains the numerous regions in the world where the political and ethnic boundaries do not coincide, yet there is no turmoil at present, such as Alsace, Switzerland, and large parts of Africa.[20] However, in other regions, ethnic groups that straddle borders want to revise them: according to Renner, "The states tear nations apart; it is not surprising that the nations want to tear the states apart."[21] The experience of various countries demands special attention, in light of the present ethnic conflicts. The Soviet Union—and Yugoslavia and Czechoslovakia—institutionalized ethnicity by the creation of ethnoterritorial administrative units as part of a federation. This created the backbone of the Soviet nationalities policy.[22] Oftentimes the borders were drawn arbitrarily in what seemed to be a policy of divide and conquer. The result was a complex and often contradictory combination of ethnic-based borders and the lack of them, as ethnic groups vied and competed against each other in the climb up the administrative status ladder.

Balkan Ethnic Entities

The Balkans are a mosaic of peoples that differentiate themselves by religion, language, culture, history and, to some extent, by biological characteristics embodied in the concept of race. If one looks at an ethnic map of Europe, one is struck by the relative homogeneity of ethnic groups within boundaries of western states and the relative heterogeneity of the eastern states, especially the Balkans.[23] Indeed, some 20 peoples coexist there, including some in the Slavic body, as well as the Hellenic, Turkik and Magyar. Furthermore, these groups practice three principal religions: Catholicism, Orthodoxy and Islam. They speak numerous languages and use three different scripts—Latin, Greek and Cyrillic. The Latin prevails in Slovenia, Croatia, Romania and Albania; the Greek script is used in Greece; while the Cyrillic is used among the Serbs, Bulgarians, Montenegrians and Macedonians.

Table 2.1 lists the size of the Balkan populations, while table 2.2 describes some demographic characteristics of Balkan peoples. At present, with the volatile interethnic situation in the region, it is crucial that there is as clear an understanding of the size of ethnic groups as the statistics can provide. Even careful scholars, in writing about interethnic issues, have all too often jumped to conclusions without an appreciation of the numbers.[24] A few words of explanation about the tables are in order. First, with respect to Albania, there are two main groups, the Tosks and the Ghegs, which are distinguished by dialect and location: the former tend to live in the south, while the latter live in the north and Kosovo. Nevertheless, they have been lumped into a single category for the purposes of this study, along with the Albanians of Kosovo and Macedonia. They are of the same ethnic stock, religion and language. (They are, of course, different from the Muslim Slavs of Bosnia-Herzegovina and Sandzak by ethnicity, although not by religion.) It is noted, however, the Albanians outside of Albania have forged a new national personality as a result of historical circumstances, mostly due to the higher levels of economic development that they have maintained since World War II. Second, Bosnians do not exist as a national group, only a territorial group: they are residents of Bosnia that are of Serb, Croat and Muslim ethnicity. The Bosnian Muslims are treated as an ethnic group despite the fact that they are identified as such by their religion. Indeed, Bosnian Muslims are racially identical to the Serbs and Croats, with whom they share a language. However, in 1945, Muslims of Yugoslavia were elevated to nation status by President Tito, thus granting them equality with Serbs, Croats and others. According to Allcock, "Precisely what is the content of Muslim identity in this ethnic sense, however, remains (deliberately?) unclear, as linguistically they are not distinguishable from either Serbs or Croats, and they are quite explicitly not people of Turkish descent, from whom they are distinguished in the census returns."[25] A similar question of definition exists with respect to the Macedonians. The present inhabitants of Macedonia, who call themselves Macedonians, are Slavs and thus not the descendants of the ancient Greek Macedonians. President Tito granted nation status to Macedonians and declared the Macedonian Orthodox Church autocephalous and thus distinct from the Serbian Orthodox Church, of which it had historically been a part. The question of whether Macedonians are a separate group or whether they are really Serbs or Bulgarians that simply speak a different language is unresolved in the minds of many. Third, the Montenegrians are treated separately from the Serbs, despite the fact that much evidence exists to support the contention that these are same ethnic groups, sharing the same religion, language and script, and differing only in historical experience. Fourth, Serbs and Croats are treated as separate groups, although in essence they differ only by religion and historical experience. Recently, Croatian scholars have attempted to prove that Croats are not Slavs, but that they are ethnically linked to Germans as they are Aryans.[26] The language Serbo-Croatian has now been divided into Serbian and Croatian, and efforts have been made in the latter to introduce words to sharpen the distinction. Fifth, although Czechs are included in the study since they populate the Balkans, Moravians have not been included since they are not significantly represented.

According to tables 2.1 and 2.2, the Balkan peoples can be observed according to demographic, territorial and religious characteristics. With respect to population size, the principal ethnic group in the Balkans are the Romanians, followed by the Greeks and the Serbs. If the (religiously, ethnically and historically) related Montenegrians are added to the Serbs, their number further increases.

With respect to ethnic spillage into their nontitular states, the Serbs are the most spread out, inhabiting sections of Croatia and Bosnia-Herzegovina. Indeed, the Serbs are second only to the Hungarians both in terms of population size and in terms of the relative size of their diaspora: there are some 11.5 million Hungarians in Hungary, 2 million in Transylvania, 350,000 in Serbia, 600,000 in Slovakia, 170,000 in Ukraine. Serbia also is home to numerous ethnic groups, including Hungarians, Muslims, Albanians and gypsies.[27] Croats live in Herzegovina and Bosnia, Albanians live in Macedonia and Greece, Macedonians live in Bulgaria, et cetera. Greeks and Slovenians are the populations most concentrated within their state boundaries.

Thus, the Balkans are a conglomeration of numerous ethnic groups, religions, scripts and languages. In the Balkans, each of these characteristics has become a political issue. Not only has ethnicity been used to withhold rights of a people, but there is not even agreement among states as to the existence of ethnic groups, as evidenced by the fact that some groups are recognized in one Balkan state only to be denied recognition in another. For example, although Macedonians have been granted national status in former Yugoslavia, they are not recognized by Bulgaria, for whom they are simply western Bulgarians (despite the fact that Bulgaria was among the first in 1992 to recognize the state of Macedonia), nor are they recognized by the Greeks, for whom they are simply Slavic Macedonians (Greeks recognize neither the nation nor the state). Indeed, when the Bulgarian and Greek prime ministers, Popov and Mitsotakis, respectively, met in March 1991, they issued a joint

Table 2.1
Ethnic/Religious Group Populations by State and Percentage of Total State Population (in parentheses)

Group	Romania (a)	Bulgaria (b)	Greece (c)	Albania	former Yugoslavia	new Yugoslavia	Croatia	Macedonia	Slovenia	Bosnia-Herzegovina
Albanians	ins.	5,000-10,000 (.09)	((95,000)) (.94)	3.1m (95)	1.73m (7.7)	1.34m (13.5)	6,006 (.13)	377,726 (20.0)	1,985 (.1)	4,396 (.1)
Bulgarians	NA	7.48m (85)	[285]	ins.	36,189 (.16)	33,479 (.34)	441 (.01)	1,984 (0.1)	105 (.1)	180 (.004)
Turks	ins.	((1m)) (8.8)	179,895 [15,725] (1.7)	ins.	101,291 (.45)	13,957 (.14)	279 (.01)	86,89 (4.5)	187 (.01)	277 (.01)
Romanians	20.4m (87.9)	NA	278	ins	54,955 (.25)	53,852 (.54)	609 (.01)	97 (.01)	94 (.01)	302 (.01)
Hungarians	((2m)) (8.6)	NA	ins.	ins.	426,866 (1.9)	390,706 (3.9)	25,439 (.55)	280 (.02)	9496 (.50)	954 (.02)
Serbians	NA	NA	NA	ins.	8.14m (37)	6.20m (63)	531,502 (12)	44,613 (2.3)	42,182 (2.2)	1.32m (32)
Croats	ins.	ins.	NA	ins.	4.43m (20)	156,272 (1.6)	3.45m (75)	3,307 (.17)	55,625 (2.9)	758,140 (18)
Muslim Slavs	ins.	ins.	NA	ins.	1.99m (9)	293,246 (3.0)	23,740 (.52)	39,555 (2.1)	13,425 (.71)	1.63m (39)
Slovenians	ins.	ins.	NA	ins.	1.75 (8)	12,570 (.13)	25,136 (5.5)	648 (.03)	1.71m (91)	2,755 (.07)
Romas (Gypsies)	((2.3m)) (9.9)	550,000 (6.2)	140,000 (1.4)	((62,000)) 10,000 (.3)	((800,000)) 168,099 (.73)	112,430 (1.1)	3,858 (.08)	47,223 (2.5)	1,435 (.08)	7,251 (4.3)
Macedonians	ins.	((187,789)) (2.1)	200,000 (1.9)	((15,000)) (.9)	1.34m (5.9)	49,864 (.50)	5,362 (.12)	1.27m (67)	3,288 (.17)	1,892 (.05)
Yugoslavs	ins.	ins.	[207]	ins.	1.22m (5.4)	473,184 (4.7)	379,057 (8.2)	14,240 (.75)	26,263 (1.4)	326,316 (7.9)

Group	Romania (a)	Bulgaria (b)	Greece (c)	Albania	former Yugo-slavia	new Yugo-slavia	Croatia	Mace-donia	Slovenia	Bosnia-Herze-govina
Monten-egrians	ins.	ins.	NA	ins.	579,043 (2,6)	547,954 (5.5)	9,818 (.21)	3,920 (.21)	3,217 (.17)	14,114 (.34)
Czechs	NA	NA	NA	ins.	19,624 (.09)	3,277 (.03)	15,061 (.33)	164 (.01)	433 (.02)	690 (.02)
Slovaks	NA	NA	NA	ins.	80,334 (.36)	73,240 (.74)	6,533 (.14)	67 (.004)	144 (.01)	80,334 (.01)
Ruthen-ians	NA	NA	NA	ins.	23,286 (.10)	19,776 (.20)	3,321 (.07)	23 (.001)	54 (.003)	111 (.003)
Vlachs	NA	((400,000)) (4.5)	((300,000)) (2.9)	10,000 (.3)	32,063 (1.4)	25,597 (.25)	16 (.000)	7,190 (0.4)	17 (.000)	49 (.000)
Italians	ins.	ins.	[2,093]	ins.	15,132 (.67)	572 (.01)	11,661 (.25)	96 (.01)	2.187 (.12)	616 (.02)
Jews	30,000 (.13)	5,000 (.05)	NA	ins.	1383 (.06)	688 (.01)	316 (.007)	27 (.001)	9 (.000)	343 (.008)
Ger-mans	280,000 (1.2)	ins.	[10,700]	ins.	8,712 (.04)	5,409 (.10)	2.175) (.05)	288 (.02)	380 (.02)	460 (.01)
Greeks	ins.	8,241 (.09)	9.49m (94)	((300,000)) or 40,000 (1.2)	1,639 (.01)	778 (.01)	100 (.002)	707 (.04)	18 (.000)	36 (.000)

Note: Double parentheses ((-)) denote a probable overestimate, followed by the official statistics, if available; ins. implies the number is insignificant; NA means that the data were unavailable.

New Yugoslavia = Serbia (with Vojvodina and Kosovo) and Montenegro

Source: Unless otherwise noted, all entries are from the following sources: Savezni Zavod Za Statistiku, *Statisticki Godisnjak Jugoslavije,* 1990, Belgrade; Hugh Poulton, *The Balkans* (London: Minority Rights Publications, 1991); *East Europe and the Republics: A Political Risk Annual,* IBC USA Publications, July 1992; John Paxton, ed., *Statesman's Yearbook,* 127th ed. (New York: St. Martin's Press, 1990-91); various RFE/RL Research Reports.

(a) Romania last published official statistics pertaining to ethnic groups in 1977. Thus the data given in this table are estimates for 1989 produced by Paxton (see above).

(b) The Bulgarian statistics refer to 1965 (the last official statistics on ethnic groups) and updates have been estimated by Poulton (see above).

(c) The official Greek statistics pertaining to ethnic groups are from 1951. The numbers in square brackets are taken from the National Statistical Service of Greece, *Statistical Yearbook of Greece,* Athens, 1992, table II.17.

Table 2.2
Demographic Characteristics of Principal Balkan Peoples

Ethnic group	Population across Balkans (approx. in mil.)	Principal home state	Other Balkan states of residence (a)	Population in state in millions and % of total	Principal religion
Albanians	5	Albania	Kosovo Macedonia	3.1 (95)	Muslim
Bulgarians	8	Bulgaria	Dobrinja Macedonia	7.5 (85)	Orthodox
Turks	1.3	Turkey	Bulgaria Bosnia	56.1 (NA)	Muslim
Romanians	21	Romania	Vojvodina Moldova	20.4 (88)	Orthodo
Magyars	2.5	Hungary	Transylvania Vojvodina	10.6 (NA)	Catholic
Serbians	9	Serbia	Bosnia Croatia	6.2 (b) (63)	Orthodox
Croatians	4.5	Croatia	Bosnia	3.5 (75)	Catholic
Muslim Slavs	2	Bosnia	Sandzak	1.6 (39)	Muslim
Slovenians	1.8	Slovenia	ins.	1.7 (91)	Catholic
Romas	4	Romania	Serbia Bulgaria	2.3 (10)	Orthodox/ Muslim/ Catholic
Montene-grians	0.6	Montenegro	Serbia	.5 (6)	Orthodox
Greeks	10	Greece	Albania	9.5 (94)	Orthodox

Note: The lower population estimates from table 2.1 were adopted in this table.

(a) Territory across states refers to the territory in other Balkan states.

(b) Refers to the new Yugoslavia.

statement denying the existence of ethnic minorities on their territories, including Macedonians.[28] Furthermore, ethnicity has been used as a political tool to increase separateness among peoples: the Slovenes and Croats are claiming that they are not of Slavic but of Teutonic tribal origins, and both Serbs and Croats are claiming that the Bogomil Slavs who became the Bosnian

Muslims belonged to their population group.[29] With respect to religion, the three principal religions of the Balkans—Catholicism, Orthodoxy and Islam—are once again pitted against each other as they were during the Turkish occupation and during World War II. Religion has been invoked in the recent past in an effort to gain sympathy for national struggles. Indeed, we have witnessed a turning to Muslim states for aid on the part of the Bosnian Muslims and a search for support from the Vatican by the Catholics in Hungary, Croatia and other regions. The Orthodox Church is less able to be a strong force since it lacks unity and is highly fragmented among the autocephalous churches of Serbia, Bulgaria, Greece, Russia and Macedonia. Scripts, too, have become politicized, as regions attempt to strengthen their traditions in a time of change by emphasizing their script. Indeed, soon after independence, the Moldavians changed their script to Roman, as it was before Cyrillic script was imposed Moscow. In Serbia, where up to a few years ago both Cyrillic and Latin script were used interchangeably, the Latin script has been eliminated from public use. And lastly, language has become a politicized issue, as language groups try to disassociate themselves. The clearest example is the attempt by the Croats to create a new language, Croatian, with the introduction of a new vocabulary. It all reminds us of the nineteenth-century statement by a Briton, "Every language must have an army."[30]

ETHNOTERRITORIAL DISPUTES IN THE BALKANS

Historical Development of Balkan Borders

Most scholars searching for the roots of the present Balkan quagmire use as a decisive date the turn of the century, with its consecutive Balkan wars and the breakup of the Austro-Hungarian and Ottoman Empires. Thus the Treaty of Versailles is a starting point of the modern period. While this is indeed a useful demarcation, the roots of Balkan territorial pretensions lie significantly farther back in history. Indeed, one must understand the peak periods in history of all the competing Balkan nations to appreciate why their current territorial and power demands are mutually exclusive and why the region can be aptly described as numerous great nations trapped in a confining territory. This historical review indicates that most Balkan nations are vying for the same territory, and this is causing problems. According to Anthony Smith, "Where there is more than one title-deed to the same territory, the probability of ethnic conflicts and nationalist wars is greatly increased."[31]

Each of the Balkan states had its period of splendor, coinciding with its maximum conquest of territory. These expansions, whether through wars or as awards by stronger powers,[32] occurred in different historical periods. In the premedieval period, Macedonia was the largest territorial entity in the Balkans. During the times of Philip II and Alexander III, the kingdom of Macedon stretched to Persia and the Mediterranean. After Alexander's death in 323, the empire fell apart and numerous ethnic groups moved into the region. During the Ottoman Empire, the region of Macedonia was limited to the Vardar Valley. This roughly coincides with the regions where Macedonians reside today. Today there exists also Greek Macedonia, which extends to the Agean Sea and includes Thessalonika, and reaches in the east toward Bulgaria, and whose western regions are inhabited with large numbers of Slavic Macedonians.[33]

It was during the premedieval times that other states, notably Slovenia and Croatia, had only been independent during premedieval times until recently.[34] This early period of independence did not coincide with their maximum territorial conquests, as shown in map 2.1. Slovenia was independent for a short time in 650 before being engulfed by the Germans, while Croatia enjoyed independence under King Tomislav in the ninth century. Under King Tomislav, the Croats established a kingdom whose territory extended north to the region between the Sava and Drava Rivers, and under King Kresimir in the eleventh century, the kingdom included Croatia and Dalmatia. In 1102, Hungarian nationals filled the Croatian throne, and Dalmatia became a part of the Venetian republic (and remained so until the eighteenth century). Indeed, its location sandwiched between the strong Venice and Hungary precluded a long independent life.

During medieval times, Bulgaria and Serbia reached their zeniths. Bulgaria, under King Simeon's rule, enjoyed its first Bulgarian Empire, which collapsed in 1018. At its peak, it included the present Black Sea coast of Romania and most of Serbia, and had an outlet in the Adriatic Sea in what is now Albania. This period was followed by years of subjection to the Byzantine and Ottoman Empires. In the nineteenth century, Bulgaria changed borders several times. After the Treaty of San Stefano (1878), Bulgarian territory included Macedonia, parts of Serbia, Eastern Rumelia, and western Thrace (thus giving it an exit to the Agean Sea). This was amended by the Congress of Berlin in the same year, according to which Bulgarian territory was greatly reduced.

Serbia covered the greatest territory during the reign of Tzar Dusan, culminating in 1355. At that time, Serbia reached the height of its glory, and its borders spread from the Adriatic to the Aegean Sea and to the gates of

Constantinople, including parts of Bulgaria and all of Albania, Montenegro, Bosnia-Herzegovina and northern Greece. The defeat came at the hands of the Ottoman Army at the Battle of Kosovo Polje in 1389. Then, during the many years of Turkish domination, Serbia was relegated to a small region with Belgrade on its northern border. Serbia achieved its autonomy from the Ottoman Turks in 1829, and it was accorded full independence in 1878 by the Treaty of Berlin. At the conclusion of the Balkan wars, Serbian territory was extended to include parts of present southern Serbia (Kosovo and Metohija) and Macedonia. Moreover, after World War I, Serbia received the predominantly Serbian Vojvodina from the Hungarian Empire.

Romania, Albania and Croatia all reached their territorial peaks in modern times. Romania, which has its roots in Rome, entered the medieval period as a series of principalities. During the Ottoman subjugation, two principalities, Wallachia and Moldavia, enjoyed some form of autonomy, in large part as a result of their distance from Constantinople. Russia, acting as a protecting state, absorbed Bessarabia in 1812. The Treaty of Adrianople called on Russia to be the protectorate of the principalities. It was only after the Crimean War (1854) that the unification of the Danubian principalities occurred and the modern Romanian state was born. Romania doubled its territory after World War I and became the largest of the Balkan states. It acquired Bessarabia on the east and the largely Hungarian province of Transylvania to the west. To the south, it had already received Dobrudja from Bulgaria in 1913. Until Stalin retook the Bessarabian territory to incorporate it into the Soviet republic of Moldavia in 1940, Romania was to remain the largest unit in the Balkans. Thus Greater Romania existed during this century, and the possibilities for its re-creation arose again in this century with the breakup of the Soviet Union and the possible incorporation of Moldavia.

Albania was created as a state in 1913, largely due to the urging of Italy and Austria. The general feeling in Europe was that its creation was neither desirable nor necessary: indeed, Bismarck said to Abdul Bey Frassari, an Albanian representative, "But you haven't even got an alphabet or a written language. How do you expect to create a State?"[35] At the time of its creation, its borders were not defined, as the British and the French instead favored the partitioning of the region among Greece, Montenegro and Serbia. This was prevented by Woodrow Wilson, leading to the recognition of the Albanian state in 1920. However, recognition did not resolve the territorial questions associated with Albania's creation. History provided few answers, since borders of what was the principality of Albania varied over time, at one point in the Middle Ages spreading south into Greece to across the Peloponnese. While under Ottoman rule, the Albanian League, a nationalist

Map 2.1

Albania and Croatia during World War II

group founded in 1878, had as its goal to unite all provinces populated by Albanians under a single autonomous state of the Turkish Empire (including Shkoder, Kosovo, Monastir and Janina). This was the first and unsuccessful attempt at creating Greater Albania. When the great powers did define Albania's borders, they gave Kosovo to Serbia and Greece got Cameria, while Albania got 35,000 ethnic Greeks.[36] This delineation persisted until the occupation of Serbia and Greece by Nazi Germany, at which time the German ally Albania was given the territories it had yearned for: Kosovo and Cameria. Under Italian tutelage, Greater Albania came to include also eastern Montenegro and western Macedonia (see Map 2.1). These borders persisted from 1941 to 1944, after which time the regions in question were reincorporated into their prewar countries. Thus World War II represented the territorial peak of Albania.

Greece, too, changed its borders over time, and discounting the period in ancient history when it flourished, it reached its territorial peak in modern times. Following the demise of the Byzantine Empire, and at the peak of Ottoman rule, Greece was subjugated in its entirety to neighboring Turks. However, Greeks did fare better than other Balkan peoples because they achieved a "senior partnership" within the Ottoman Empire.[37] Wars of independence led to Greek autonomy in 1829 with the Treaty of Adrianople (the same treaty that confirmed Serbian autonomy and placed the Romanian principalities under Russian protectorship). At this time, the territory of this new Greece only included one-fourth of the Greek people of the Balkans, and its borders came to slightly north of Athens. After World War I and the Balkan wars, despite the fact that Greece was aligned with the victors, it did not receive as much compensation as requested. While the Bulgarian parts of Thrace were awarded to Greece, Britain retained Cyprus and Italy retained the Dodecanese Islands. Greece did receive Smyrna (Izmir) and the surrounding territory that contained a significant Greek population. However, in 1921 war broke out in the surrounding territory of Anatolia between the Greeks and the rejuvenated Turks under Ataturk. Turkish victory resulted in the loss of Greek rule in Smyrna and the loss of parts of Thrace and two islands.[38]

Like Albania, Croatia also reached its territorial peak during World War II. After the takeover by Hungary in the eleventh century, Croatia became a part of the Hungarian Kingdom, then became a part of Austrian Empire, and then in 1867 was once again turned over to the Hungarian administration. At the end of World War I, Croatia, as a losing nation, was united with winning Serbia to form the Kingdom of the Serbs, Croats and Slovenes. However, during Austrian rule, the present territory of Croatia included Croatia proper, Dalmatia, Slavonia and Krajina. These were separate administrative regions included under the administration of Zagreb, with the exception of Krajina, which was ruled directly from Vienna until 1867. Croatia voluntarily decided to join the Yugoslav union, as years of pan-Slavic nationalism had permeated the Yugoslav lands and instilled the strong desire to unite all south Slavs into one state. Disappointment with the union and nostalgia about the Germanic links during the interwar period led to the rise of a separatist movement among the Croats, who in World War II opted for cooperation with Hitler and the establishment of an independent state. Indeed, Hitler's expansion in the Balkans gave the Croats not only independence, but also an enlarged territory. After 1941, Croatia included Bosnia-Herzegovina and parts of Montenegro and it extended

Map 2.2
Internal Boundaries of Yoguslavia, 1929 (*Banovine* System)

into Serbia as far as Belgrade. In the aftermath of World War II, when Croatia reentered Yugoslavia, it benefited insofar as it was awarded almost all of the Istrian peninsula, former Italian towns in Dalmatia and some islands in the Adriatic.[39]

No discussion of internal Yugoslav boundaries is complete without a mention of the system of *banovine* (Map 2.2). During the interwar period, two different administrative divisions existed, composed first of 33 administrative units, then after 1929, of 10 (9 of which came to be called *banovine*, plus the City of Belgrade). These units were an innovation insofar as they did not correspond to ethnic divisions; instead they represented geographical

Table 2.3
Administrative Territories of Yugoslav Regions—pre-1914, Interwar and Communist Periods

Pre-1914 Regions	*Banovine* (1929-1938)	Republics (1946-1991)
Slovenia	Drava	Slovenia
Dalmatia, Slavonia, Croatia	Sava, Coastal, Zeta	Croatia
Bosnia-Herzegovina	Vrbas, Drina, Zeta	Bosnia-Herzegovina
Vojvodina, Northern Serbia	Morava, Dunav, Drina, City of Belgrade	Serbia
Montenegro	Zeta	Montenegro
Southern Serbia	Vardar	Macedonia

configurations (each banovina carried the name of its principal river). Such a regional administrative system may have achieved the goal of decreasing the importance of ethnicity, but the pressure from the Croats was too strong. In 1939, the central government ceded to Croatian pressure to form the *banovina* of Croatia, which included the Coastal and Drava *banovine*, as well as parts of Dunav, Vrbas, Zeta and Drina. Table 2.3 contains a rough correspondence of the regions in three time periods.

The system of *banovine* was dropped after World War II and replaced by a federal system according to which the administrative boundaries only partially coincided with ethnic boundaries. Such map drawing was deficient, since some ethnic groups were awarded regions while others were not. Indeed, while Slovenes became concentrated in Slovenia, Montenegrians in Montenegro, and Macedonians in Macedonia, the Serbs and the Croats were not all enveloped in their republics. Croats were left out of the demarcations of Croatia, mostly in Herzegovina and Montenegro. Large numbers of Serbs found themselves in the neighboring republics of Croatia and Bosnia-Herzegovina. At the same time, two autonomous republics were carved out of Serbia, on the grounds of satisfying the minorities residing there. While this may have been valid in Kosovo, where over half of the population in 1951 was Albanian, the argument certainly did not hold in Vojvodina, where less than 20 percent were Hungarian. The fact that similar status was not extended to the Croats of Herzegovina and the Serbs of Croatia highlighted the inconsistency that was the chief characteristic of the federal internal boundaries of Yugoslavia.

Current Ethnoterritorial Disputes

The borders of the Balkans have evolved to their present form over history as a result of geography, conquests and international intervention, rather than along strict ethnic lines. Thus, in the current revival of ethnicity, it is not surprising that the region is experiencing numerous ethnoterritorial disputes. These are presented below. While these disputes differ in numerous characteristics, such as their intensity, their manifestation, their duration and their historical roots, they have all surfaced or resurfaced in the course of the early 1990s and are ethnically based (with the possible exception of the conflict in Istria, which seems to be regional in nature).

Istria

This peninsula, presently divided between Slovenia and Croatia, has at various times been part of the Venetian Republic, Italy, Austria-Hungary and Yugoslavia. Today, Istria hosts two separate ethnoterritorial conflicts. The first is between Slovenia and Croatia, whose relations became strained because of mutual disappointment at each other's lack of aid during their respective wars of secession. Even two years after the fact, verbal antagonism reveals the betrayal that both sides perceive. This is coupled with the economic concerns each has over assets located on each other's territory that to date have not been resolved.[40] However, their greatest source of antagonism has to do with borders: President Tito's borders left some Slovenes and Croats on the wrong sides and left Slovenia with what it considers unsatisfactory access to the sea routes. Thus the question of the borders in Istria, as well as the maritime borders of Slovenia and Croatia, is unresolved.[41] Indeed, tensions came to a head in May 1993 when Croatia set up a new border post in a region that Slovenia claims.[42] The net result of these disagreements is that according to a poll taken in Slovenia in 1992, 46.5 percent perceived Croatia as a state that could endanger the independence of Slovenia (whereas in 1991, this number was only 3.5 percent).[43]

The other territorial issue in Istria involves Italy. Italians resided in Istria and other parts of Dalmatia before World War II. The Italian minority that was displaced after World War II is demanding, in the least, compensation for confiscated property, and at the most, a restoration of prewar boundaries with Italy.[44] The basis of the Italian argument is that the Treaty of Ossimo, defining the borders between Yugoslavia and Italy, was drawn up with Yugoslavia, but since Yugoslavia no longer exists, the terms of the treaty, including borders, should be up for renegotiation.[45] This unresolved issue has caused instability in the Italian government, and its resolution is the focus of intergovernmental discussions.

Recently, two developments have taken place that will affect these territorial issues. First, with the Italians, progress is being made on peaceful negotiations. In January 1993, Croatian premier Sarinic and Italian president Scalfaro agreed to reciprocity in minority rights of residents in each other's states, and most important, Croatia agreed to absorb Yugoslavia's commitment to pay $94 million in compensation to Italian refugees after World War II.[46] Second, the past year has witnessed the rising popularity of a regional political party, the Istrian Democratic Alliance, that supports the creation of an autonomous Istria within Croatia.[47] This party won 72 percent of the local elections held in February 1993, forming the strongest opposition to the nationalist party of President Tudjman, the HDZ. While its platform is not secessionist, it is demanding autonomy within Croatia. That autonomy includes greater financial distance from Zagreb, based mostly on the estimated 90 percent of income generated in the region that is siphoned off to the center. Tourism generated some $500 million to $700 million in 1993, and the regional population resents having to share this wealth.[48] This rise in popularity of the Istrian autonomist party has been cooly received in Zagreb, and there is evidence that the Zagreb government is sponsoring the settlement of Croatian refugees from the war in Krajina and Bosnia in order to dilute the Istrian population.[49]

Krajina and Slavonia

The region of Croatia where the Yugoslav war was fought in 1991 is known as the Military Border (Vojna Krajina, or Krajina for short). This region was settled, mostly in the mid-sixteenth century, by Serb migrants that took refuge from the Turks. These were military-style colonies whose inhabitants had special privileges in return for protection of the Hapsburg lands. Thus Serbs of the region got the name "Guardians of the Gate." This region was ruled directly by Austria (even after the Ausgleich of 1867 assigned Croatia and Slavonia to Hungarian rule) until 1881, when it reverted to Croatian rule under a type of autonomous status. During World War I, when Austria-Hungary went to war against Serbia, the Serbs of the Military Frontier were called upon to fight their kin.[50] During the interwar period, the Krajina region was divided among several *banovine*, and after World War II, it was overwhelmingly allocated to Croatia, with a portion given to Bosnia-Herzegovina. The question that motivated the civil war of 1991 was the status of Krajina.

The Croats claim that Krajina is an integral part of their land because it has been under the rule of Zagreb, in some form or other, since 1881. The Serbs in Krajina claim that they won't succumb to rule by Zagreb, but will exercise their right to self-determination (as Croatia did, after all, when it left the Yugoslav union), since they form the majority of the population (in most districts)[51] and have resided there for centuries. While Croat nationalism thrived in Croatia and in the Croatian diaspora in the late 1980s and 1990, an analogous sentiment fermented among the Serbs of Krajina and Slavonia. They sensed the upcoming possible secession of Croatia from the Yugoslav federation and expressed refusal to be part of an independent Croatia. This sentiment was due to their experience during World War II, when they were terrorized and slaughtered by the fascist Ustashas. Nationalism prevented Croatian president Tudjman from issuing assurances to the Serb population, and nationalism swayed the Serbian population to favor armed efforts to protect itself. They were aided by Serbian president Milosevic, who not only assured the Krajina Serbs of his support, but also abetted their upheaval. On the basis of their own referendum, the Serbs of Krajina voted overwhelmingly in favor of seceding from Croatia *if* Croatia seceded from Yugoslavia and subsequently established the self-proclaimed Serbian Republic of Krajina. War broke out in June 1991, and the former Yugoslav army came to the aid of the Krajina Serbs. There was bloodshed and property damage, and refugees were created on both sides. At the time of this writing, the UN peacekeepers are trying to control the area through a complicated system of pink and blue zones, despite periodic flare-ups such as those around the Maslenica Bridge (spring 1993) and Karlovac (September 1993).

Thus it is too late to discuss the way in which Croatia should have seceded, or whether the world should have granted it recognition when it did. The question on the table presently is where the borders of Croatia should be drawn: should they reflect the ethnic composition and historical territorial rights of the Serbs of Krajina and Slavonia, or should they reflect the administrative borders drawn up by President Tito?

Vojvodina

Vojvodina is a territory including Backa, Srem and the Banat, inhabited both presently and historically by a large number of ethnic groups, including Serbs, Hungarians, Romanians, Ruthenes and Ukrainians, Germans, Slovaks and Croatians. This territory was for centuries part of the Austrian Empire, and like Krajina in Croatia, it formed part of the Military Border where Serbs enjoyed special status and privileges. Part of it was the original Serbian Vojvody, and by the mid-1850s, the Serb population agitated for the restoration of this historical autonomy and unification with Serbia.[52] In 1867,

Hungary took control of Vojvodina, and in World War I, the region was granted mostly to Serbia and partly to Romania (eastern Banat). Vojvodina's status within the Yugoslav federation was that of an autonomous region within Serbia, a status greatly resented by Serbs because of the absolute majority of Serbs in the population and the fact that it was Serbian military victory in World War I that got Yugoslavia the region. This status was created ostensibly to satisfy the Hungarian population, despite the fact that Hungarians after World War II amounted to under 20 percent of the population and were concentrated in some eight districts along the Hungarian border.[53] According to Reisch, the Hungarians living in Vojvodina "benefitted from the best treatment of all the minorities in neighboring East European countries."[54] This autonomous status persisted until 1990, when the Serbian government under Milosevic revoked the powers and status that were granted Vojvodina and Kosovo under the 1974 constitution and thereby reincorporated it into Serbia.

Since the reincorporation of Vojvodina into Serbia, relations between Belgrade and the Hungarian minority in Vojvodina have deteriorated, as have those between Hungary and Belgrade. With the "spiritual support" of the Hungarian prime minister Jozsef Antall and under the auspices of the Democratic Union of Hungarians in Vojvodina, they are demanding autonomy and self-government in a three-tier program for personal rights, local autonomy and territorial autonomy.[55] It is estimated that some 50,000 Hungarians have left for Hungary, due to fear that Serbian nationalists would pressure them into leaving.[56] However, while the Hungarian minority in Vojvodina is concerned with its rights, safety and the possibility of conscription into the Yugoslav army, there is equal fear among the Serbs of Vojvodina that the Hungarian government in Budapest may make use of the present Balkan instability to reaffirm or regain control of territories it lost after World War I. This feeling is strengthened by declarations such as that of the Hungarian prime minister, who noted that the territories his country lost to Yugoslavia following the Paris Peace Conference (1919-1920) were ceded to a country that no longer exists, and thus this cession should be reevaluated.[57]

Bosnia-Herzegovina

At the time of this writing, Bosnia-Herzegovina is the scene of a gruesome civil war, in which three peoples (the Bosnian Serbs, the Bosnian Croats and the Bosnian Muslims) are battling each other for their lives and their property. Despite the efforts of some players in this war to internationalize the conflict,[58] this is a civil war between three ethnic/religious groups that, driven by nationalism and the desire to protect themselves and their property, fight for territory in which to proclaim their nation-states.[59] That there really are three sides is clearly indicated by the existence of three different political parties, each one associated with one ethnic/religious group.[60] Two sides, namely Bosnian Croats and Bosnian Serbs, are irredentist in nature and desire ultimate union with Croatia and Serbia respectively.

The war broke out in response to the international recognition of a sovereign state of Bosnia-Herzegovina. The Serbian population of Bosnia, which amounted to some 32 percent of the population and had property rights to some 60 percent of the territory, boycotted the secession referendum and started taking steps to ensure that, in the event of Bosnia's secession from Yugoslavia, they and their territory would not leave the Yugoslav union. Thus, the logic behind the beginning of the hostilities was the same as in Croatia insofar as one ethnic group demanded the right to exert its right to self-determination just like others. Again, as in Croatia, the dispute between the sides was aggravated by the international community, not because it granted recognition, but rather because it did so without consideration of the desires of the Serbs. Indeed, the Lisbon accord, drawn up and signed by the three warring sides in March 1992, was similar to the plan on the table in September 1993, except that it was more favorable for the Muslims than later plans.[61] However, President Izetbegovic, strengthened by the encouragement of the United States to seek a unitary state, rescinded his signature, and the war began in earnest.

The ensuing civil war between Bosnian Serbs, Bosnian Croats and Bosnian Muslims resulted in a vicious battle, loss of lives, loss of property and loss of the ideal of interethnic harmony. It also resulted in the proclamation of two self-declared regions: the Serbian Republic (in spring 1992) and the Croatian Community of Herceg-Bosna (in summer 1992, and elevated to Republic status in summer 1993). Then, in September 1993, Fikdar Abdic proclaimed the Autonomous Province of Western Bosnia, centered in the Bihac area, and thus seceded from the Muslim territories of Bosnia. In the spring of 1993, the ongoing peace conference in Geneva almost achieved the division of Bosnia-Herzegovina into eight to ten ethnically based regions with a loose central government. This Vance-Owen Peace Plan, named after UN mediator Cyrus Vance and EC mediator David Owen, was replete with problems since its inception. First, the map contained a series of Nagorno-Karabakhs, or unconnected ethnic enclaves, that were unviable economically or politically. Second, the plan called for a strong unitary central government, which satisfied only the Muslim side (some 40 percent of the population). The Croats and the Serbs were against such a political form, opting instead for a loose confederation of republics.[62] Third, the plan would have been agreed to under severe international pressure and thus would have had to entail substantial foreign support in the form of troops, infrastructure and capital to create and then maintain the peace. None of these were forthcoming from the West, already exhausted from its failed and unpopular efforts in this Balkan crisis, and offers from the Muslim countries were rejected in order not to increase Muslim influence in Europe. For all these reasons, the plan had no hope of bringing about stability in the region. For the sake of ending the misery of so many people, the international community should efficiently and quickly agree to drawing the lines among the warring groups. Indeed, as Hayden pointed out, when partition lines are drawn by decree (such as in Punjab) they are less painful, and the adjustment is quicker, than when they are drawn by force (as in the case of Bosnia).[63]

At the time of this writing, the plan that is under negotiation features the following: (1) a stipulation that after two years, each of the republics of Bosnia-Herzegovina will have the right to a referendum to decide whether to secede, and (2) a map according to which the Serbs get 52 percent of the territory, Muslims get 30 percent and the Croats get 18 percent. Muslims are to get an outlet to the sea at Ploce on the delta of the Neretva River in the form of a 99-year lease on docks and surrounding land (moving traffic cannot be controlled by Croat authorities). Whether these points will be accepted without modification is at this point not clear.

Thus it is proving a difficult task to create a state called Bosnia-Herzegovina.[64] As pointed out by Conor Cruise O'Brien, it "has never been a state in its history"[65] and this fact must underlie all international efforts to create one. Indeed, Bosnia-Herzegovina was part of Serbia; it was part of the Ottoman Empire; it was annexed by Austria; it was part of Yugoslavia; and it was part of Independent (Fascist) Croatia during World War II. Why does the international community think it can create a multiethnic state where one never existed before, when Yugoslavia as a multiethnic creation did not succeed.[66] As long as mother states exist to inspire both the Serbs and the Croats, it is unlikely that the majority of the Bosnian populations will be happy with such a solution.

Romania-Moldova

The region of Bessarabia has for two centuries been the focus of strain between Romanians and Russians. In the early 1800s, Turkey dominated the principalities of Moldavia and Wallachia. Moldavia reached from the right bank of the Dniester River to across the Prut River as far as the Carpathans.

In 1812, Russia annexed the region between the rivers (i.e., Bessarabia), while the rest remained under Turkish rule until 1878, when it became part of the independent state of Romania. After World War I, Bessarabia was integrated into Romania, a move that was never recognized by the Soviet Union. This led to the ceding of Bessarabia (and northern Bukovina) to the Soviet Union in mid-1940 and the subsequent return of both regions to Romania in 1941 (Romania was Germany's ally).[67] After World War II, the regions were integrated into the Soviet Republic of Moldavia.

Despite the formal recognition in 1947 by the Romanian government of the Soviet integration of Bessarabia, the Bessarabian question has remained a sore point in Soviet-Romanian relations. Ethnic Romanians made up 65 percent of the population[68] and refused to be Russified in the decades under Soviet rule. They retained their language and in the turmoil of 1989 expressed interest in integration with Romania. Then, when the Turkish-speaking Gagauzi expressed a desire to secede and the Russians west of the Dniester agitated for an independent state, the Moldovians were left with a dilemma as to what path and which alliance to pursue. Despite immediate moves to revert to the Latin script and increased requirements for the use of Romanian language in public life, support for reunification with Romania was not widespread, according to polls in the newly renamed Moldova.[69] Similarly, the sentiment in Romania is one of ambivalence. The voice favoring unification with Moldova is weaker than that opposing, and is understandably stronger in the western regions of the country, around Jassy.

The events leading up to the possible unification were tumultuous. Several Moldovian members of parliament are in favor of a gradual unification with Romania, while President Snegur is opposed. He is calling for a referendum of the people to decide (in hopes of blocking unification), which as of summer 1993 had not yet occurred. Romania has also asked the neighboring state to call off plans for the referendum, claiming that it was inappropriate at the time.[70] Instead, Romania has proposed a "treaty of fraternity and integration" with the Romanian-speaking majority of Moldova, which was turned down. Other Romanian overtures are economic in nature: it has offered economic assistance (in August 1993) if Moldova does not ratify participation in the CIS. Incidentally, this was to be financed by the Fund for Integration with Moldova, and it was to offer Moldova a lucrative alternative to economic association in the CIS.[71] Ultimately, a less ambitious economic package was signed in September 1993. Also, the chairman of the Parliament, Petru Lucinschi, who is overwhelmingly popular, is leaning toward an eastern orientation for Moldova rather than a western one. These are all setbacks for the pro-unification lobby. In addition, at this time President Snegur is more concerned with events on the eastern side of the Dniester River, where the largely Russian and Ukrainian population proclaimed the Trans-Dneistrian Republic and formally seceded from Moldova in September 1990.[72] This act, which occurred in retaliation for the nationalist push for reunification with Romania and the accompanying legislation that heckled non-Romanians-making Romanian the state language,[73] re-Latinizing the alphabet and adopting a flag and state seal very similar to those of Romania. The situation was further aggravated when, in September 1993 General Aleksandr Lebed was elected as deputy to the Supreme Soviet of the "Dniester Republic" on a platform calling for unification with Russia. For these reasons, the demands for joining Romania have subsided as the government is focusing on efforts to calm the Trans-Dniestrians.[74]

In a world in which two of Romania's neighbors are disintegrating, it is not surprising that multiethnic Romania would feel strain. In addition to the question of Moldova, there is another territorial dispute with the Ukraine, pertaining to the Serpents Island in the Black Sea. This island is rich in undersea oil and gas, was taken by the Soviet Union in the 1940s and has been publicly claimed by Romania since 1991.[75] Furthermore, given Romania's Hungarian population and its increased restiveness and assertiveness, an incorporation of some two million Romanians into Romania would unequivocally shift the demographic balance even further to the advantage of the Romanians.

Transylvania

The current Romanian territory consists of territory gained after World War I: Transylvania (previously under Hungary), Bukovina (previously under Austria), Bessarabia (previously under Russia) and southern Dobruja (previously under Bulgarian rule). In 1940, Hungary recovered its lost territory of Transylvania, only to lose it again at the conclusion of World War II. While under Hungarian rule, the Romanian population of Transylvania considered itself distinct from the other two ethnic groups—the Hungarians and the Germans—and desired political expression of that sentiment. Indeed, minorities had greater equality when Transylvania was ruled by Austria than when it became part of Hungary (1867-1918) and its parliament was abolished and self-governing status repealed.[76] Romanians did not forget the treatment that they received at the hands of the Hungarians, and they proceeded to return in kind. Today, Transylvania contains a large Hungarian population (approximately two million; 40 percent of the total in the region) that, despite efforts at appeasement, has been restive since Ceausescu's demise. These Hungarians never forgave the Treaty of Trianon, which left two thirds of Hungarian territory and population outside Hungarian borders. Ethnic Hungarians were insulted in 1965 when their feeble autonomy status was removed, administrative regions were drawn that failed to coincide with ethnic concentrations, and the policy of demographic dispersal brought numerous non-Hungarians onto traditionally Hungarian territory.

Interethnic clashes occurred in Tirgu-Mures in March 1990, as people vented anger in the vacuum created by the death of Ceausescu. The political changes in 1989 gave rise to a political party aimed at protecting Hungarian interests, the Democratic Alliance of Hungarians in Romania. While it presently is concerned with the restoration of minority rights and does not demand outright secession, irredentist sentiment is growing in Hungary, and Romania is expressing fears for the territory in which it invested heavily over the communist years.[77] Indeed, the public expressions of the Hungarian government with respect to the Hungarians in Vojvodina also apply to Transylvania. Adding to this is the emergence of a plan for cooperation that would unite the Carpathian region, which has elicited a suspicion from President Iliescu that the next step will be the inclusion of Transylvania.[78] At the same time, Romanian nationalism has increased, as Romanians are venting their frustration about the Hungarians as well as other minorities; this trend is reflected in the swift rise of the Vatra Romanaesca movement. This movement (and its political arm, the Party of Romanian National Unity) stands against cultural pluralism and concessions to minorities and in favor of a dominance of Romanian nationalist sentiment. In some regions of the country, there is widespread support for this movement. Even President Iliescu stated that the Vatra has been "an element of stability and equilibrium for the country."[79] According to Gallager, the Hungarian and Romanian movements have reinforced each other, although "it is doubtful whether the Hungarians realized that by creating their movement so rapidly, . . . they would be converting a disquieting message to a large number of Romanians."[80]

Macedonia

The importance of Macedonia is indicated by its location in the heart of the Balkans. Roucek claims that "whoever dominates the Vardar Valley is master of the Peninsula."[81] The possession of Macedonia has been the strategic ambition of three Balkan powers: Bulgaria, Serbia and Greece.

Yugoslavia was good for Macedonia, not only in terms of the capital infusions it received from the federation, but also because its separate identity was created by President Tito: Macedonians were given a language, a national identity and a church. When Slovenia and Croatia seceded, Macedonians feared that they would be subject to hegemony of a larger Serbian majority within the federation. Thus the Internal Macedonian Revolutionary Organization (IMRO), the interwar pro-Bulgarian and anti-Yugoslav party, was revived, in the form of the Democratic Party for Macedonian National Unity. In September 1991, Macedonia held a plebiscite on sovereignty, indicating that despite the problems associated with independence, the region wanted to hop on the bandwagon with Slovenia and Croatia. When the European Community set forth guidelines that it would follow for the recognition of former Yugoslav republics, Macedonia (and Slovenia) satisfied the requirements, including the one about fair treatment of minorities. Indeed, there has been harmony between the Albanians, Serbs and Macedonians. In addition, there is a Roma population that has been called "a success story": "the Roma of Macedonia appear to enjoy a far more advantageous situation than do their counterparts in Greece, Bulgaria or Romania."[82] However, scratching below the surface reveals another dimension to the interethnic situation. There have been some incidents that have dampened the relative interethnic harmony. For example, there were violent anti-Muslim demonstrations in Skoplje in February 1993; there are reports that Albanians want to declare a western Macedonian autonomous republic;[83] the main Macedonian Roma parties are requesting the creation of a state for the Romas (Romanistan);[84] and there is the fact that the Macedonian constitution does not accord non-Macedonians the same status as it does to Macedonians.

Probably the most visible issue pertaining to Macedonian independence has been its effect on Greece, where it has provoked a nationalist backlash and become a rallying cry.[85] Like Bulgaria, Greece negates the ethnicity of the Macedonians, but unlike Bulgaria, Greece does not even recognize the political entity pending the changing of the state's name. Greece is presently apprehensive about the creation of a Macedonian state to the north, especially since it perceives that such a state will have territorial claims on northern Greece (Greek Macedonia). This Greek sentiment is partially agitated by the overt use of Hellenic symbols by the Macedonian government, including on the flag and currency note. Such infringements are perceived by Greece as inappropriate and as evidence of territorial claims against Greek land. In addition, Greek suspicions of the motivations of Macedonians have their origins in the role that Yugoslav Macedonia played before and during the Greek civil war, when it supported the Greek communists. Also, Georgievski, the leader of the Macedonian independence party (IMRO) said in 1991, "Some 250,000

Macedonians live in Greece, and 51 percent of Macedonia's territory lies there. . . . A united Macedonia has long been the dream of Macedonians. . . . We want to achieve this unity peacefully."[86] This issue has created a frenzy among the Greek population both domestically and in the diaspora, leading to a blockade of the northern border with Macedonia aimed at applying political and economic pressure. It has fueled the opposition to the ruling New Democracy Party, as both the Political Spring Party and PASOK have made it an issue. Indeed, contender Andreas Papandreou has been quoted as saying, "The name Macedonia is our soul."[87] A compromise was reached when in April 1993 the new country was accepted into the United Nations as the Former Yugoslav Republic of Macedonia. This is a temporary name, as negotiations proceed for a future name.

The traditional Bulgarian demands on Macedonia seem to have been stronger during the lifetime of Tito than at present. In fact, Sofia rushed to recognize Macedonian independence, although it still negates the existence of a Macedonian ethnicity, preferring to refer to the inhabitants of Macedonia as western Bulgars. While Bulgaria officially seems to have no desire to participate in a carving up of Macedonia, there is evidence that such a sentiment is not shared by its population.

The least belligerent group at this time is the Serbian population of Macedonia. It amounts to a mere 2.4 percent of the population and is located in several districts on the northern border with Serbia. Despite the attention of the media on this group as a potential cause of the spillover of the Bosnian crisis into Macedonia, there is little evidence that such a move will come. Indeed, it is much likelier that a crisis will occur as a result of irredentist aspirations of the Albanian population. The demands of the Albanian population in Kosovo are closely related to those of the Albanian population in neighboring Macedonia, where 20 to 40 percent of the population is ethnically Albanian.[88] Their demands include, at the most, secession from Macedonia (and the fusion with Kosovo and possibly Albania), and in the least, greater autonomy and cultural rights within Macedonia (see the section on Kosovo).

Sandzak

Sandzak is a territory that lies north of Kosovo, between Serbia and Montenegro; it was known as Raska during the medieval Serbian kingdom. It was under Ottoman rule until 1878, at which time the Congress of Berlin awarded the Austro-Hungarian Empire garrison rights (not administrative rights) over it. Although this was done in part to prevent Serbia and Montenegro from uniting, the Serbian Army nevertheless won the territory from the Turks in the First Balkan War. It was divided between Serbia and Montenegro and continued as such under Titoist administrative borders.

The proportion of Muslims in Sandzak differs in the Serbian and Montenegrian regions: in the Serbian Sandjak, 60.5 percent of the population is Muslim, whereas in the Montenegrian Sandzak, 40 percent is Muslim.[89] The combined population is 52 percent Muslim, with strong links to the Bosnian Muslims. While this population is concentrated in only three districts, it is nevertheless demanding autonomy from Serbia for the entire Sandzak region, with possible integration with a future Muslim Bosnia. These demands are made under the auspices of Stranka Demokratske Akcije, a party that identifies itself as a Muslim party that openly supports the Muslim actions in both Kosovo and Bosnia-Herzegovina. Given the events of the past two years with respect to sensitivity about minorities, it is understandable that the Muslims of this region fear Serbian repression. On the other hand, given the association with the Turks that the Sandzak Muslims have expressed, and the symbols that they have chosen (such as the use of the Turkish flag as their own), it is understandable that the Serbs in the region feel nervous.[90]

Kosovo

Ramet refers to Kosovo as the "Serbian Jerusalem," Zametica as "the desperation of Serbia."[91] To Serbs, Kosovo represents the epicenter of their culture and history and the whole essence of what it means to be a Serb. Historically, the roots of this national obsession lie in the last battle that was fought unsuccessfully against the Ottoman Empire, which resulted in five centuries of domination by the Turks. The mythology created around the Battle in Kosovo Field was passed from generation to generation and fueled the rebellion that ultimately led to the independence of the Serbs from the Turks. As a result, Serbs in Serbia and the diaspora feel that its future is not negotiable, making this probably the only issue on which the majority of Serbs agree.

However, this historical cradle of Serbian civilization is presently inhabited mostly by ethnic Albanians. Due to a variety of reasons including high birth rates, in-migration by Albanians and out-migration by Serbs, the demographic picture has changed to the degree that today, some 85 percent of the population is Albanian. Indeed, the Albanians of Kosovo have the highest birth rate in Europe, resulting in the youngest population structure and thus a large dependency ratio. There has been significant in-migration of Albanians since World War II, peaking during the 1960s. Serbs, responding to both political push factors (harassment by the Albanian majority) and economic pull factors (the opportunities for economic improvement elsewhere), have migrated in large numbers. It is estimated that between 200,000 and 300,000 had left by the mid-1980s.[92]

These demographic alterations did not occur in a vacuum. The Albanians of Kosovo were given the opportunity to embark upon nation-building, mostly in 1974, when the new constitution elevated the region of Kosovo to an autonomous republic, granting the region's ethnic Albanians their own press, university, television stations and so on. The constitutional description granted the Albanians veto power in the Serbian parliament, but this right was not reciprocated. At this time, Serbs fled the region; Zametica called it ethnic cleansing.[93] The injustice the Serbs felt, coupled with the lack of response to events by both the communist and international media, led to pent-up resentment that found a voice with the emergence of a new political figure, Slobodan Milosevic. He proceeded, with the overwhelming support of the Serbian population, to alter the Serbian constitution and abolish the former autonomous status of Kosovo, and then in July 1990, he dissolved the Kosovo government and assembly and increased the police force in the region. The Albanian leadership responded to this mounting evidence of repression first by rioting, then by proclaiming its independence from Serbia and boycotting all elections and political involvement. The government in Belgrade is presently applying extensive authoritarian measures to quell the Albanian silent rebellion.

The present conflict between Serbs and Albanians in Kosovo rests on the Serbs' view that this is traditionally their territory and that the demographic imbalance that now disfavors them is due to unjust circumstances. The Albanian response is that the Albanians are the majority ethnic group in the region and they have the right, on the basis of self-determination by ethnicity, to their own state. Under the leadership of Ibrahim Rugova, Albanians in Kosovo are demanding, in the least, restoration of Kosovo's pre-1989 autonomy status, and at the most, full independence, possibly to pave the way for eventual incorporation into Greater Albania.[94] However, in May 1993 Kosovo "prime minister" Bukosia proposed a new eight-point plan according to which Kosovo would be placed under international protection and the Albanians' demand for self-determination would possibly be dropped.[95] Another possibility is the partition of Kosovo so that the Serbs get the land containing their medieval monuments. However, this idea is complicated by the fact that that land is not immediately adjacent to Serbia.

Any efforts at the re-creation of Greater Albania raise the hackles of Macedonia and Greece, and bring into question the stability of the entire region. The relationship between Kosovo and Albania has been evolving and is presently taking an interesting turn. Indeed, the official Albanian position until fall 1992 was that Kosovo is an integral part of Albania. Since then, Tirana began supporting autonomy rather than full independence. However, such a feeling is not shared by Rugova, the president of the self-proclaimed Kosovo republic (recognized only by Albania). The Albanian nationalists claim that Greater Albania should extend beyond Kosovo as far as Skoplje and Veles in Macedonia, south of Yannina in Greece, and north into regions of Montenegro. Irredentist sentiment has also spread among the Albanians in Macedonia, where the demands on the young Macedonian democracy are pushing the new government of Kiro Gligorov to the limit: in addition to use of their language, the Albanians are insisting on using their symbol, the national flag of Albania, which the Macedonians view as a provocation.[96] However, this is not to say that Albanians are not searching for a creative solution. The Albanian president Sali Berisha has actively approached the Russians for support of Kosovars, on the grounds that the Russians created the Volga Republic for Germans as a way of satisfying ethnic rights of Germans, and thus the Serbs should learn from them.[97]

Dubrovnik City-State and Autonomous Dalmatia

While the Dalmatian coast changed rulers often during history—it has been under Austrian, Ottoman and Italian rule, with a period under Napoleon's rule as part of the Illyrian Republic—the city-state of Ragusa, Dubrovnik's former name, had an independent history.

The reestablishment of the Dubrovnik Republic has been the desire of some of its residents several times in this century. Most recently, this movement was reborn during 1989-1990 under the leadership of Aleksandar Apolonio, the president of the Board of the Movement for the Autonomy of Dubrovnik. He and his movement have been under criticism from the Croatian government and press, and their efforts have culminated in a highly publicized trial of six individuals.[98] Apolonio claims that the idea for the reestablishment of the Dubrovnik Republic did not suddenly reemerge, but rather that it never died in the hearts and minds of the population. Indeed, history separates them from the Croats, and it was not until the modern era that Dalmatia came under the rule of Zagreb.

The movement for the establishment of a Dubrovnik republic must be seen against the background of the simultaneous emergence of a movement for an autonomous Dalmatia. Indeed, according to polls, the Dalmatian Action party (DA) ranks second in Dalmatia, behind the Liberals and ahead of the ruling HDZ. President Tudjman has called this party as well as the Istrian autonomy party, an enemy of the state.[99] Moreover, members of the DA party have been arrested for alleged acts of terrorism, and its leader was evacuated from her home in October 1993.[100]

Cyprus

The ages-old conflict between Turkey and Greece is rekindling, as Greece has become increasingly nervous in the post-cold war period: Albania and Turkey are establishing strong economic and political ties, and the Muslims of Bosnia hold the sympathy of the international community. These events have reminded the Greek population that their conflict with the Turks in Cyprus, while dormant, is by no means over. Therefore, the Cyprus conflict has once again been made into a major issue of domestic politics. (Indeed, the new Political Spring party has accorded Cyprus importance, and its leader, Samaris made travel there a priority early on.)[101]

Historically, Cyprus was inhabited mostly by Greeks until the Ottoman invasion in the sixteenth century. In 1878, the British assumed control of the island and remained there until 1960, when, under pressure of the Greek independence movement, it gained independence. However, one-fifth of the population (720,000 people) is Turkish, and this population refused to live under Greek proportional control in politics. A coup in 1974 led to the uprising of the Turks, aided by the army from the mainland, and the establishment of the current status quo; that is, one-third of the country is currently known as the Turkish Republic of Northern Cyprus (formally established in 1983 and recognized by only Turkey), and the Greek section is the Republic of Cyprus. Since 1974, UN peacekeeping troops have been maintaining peace, and negotiations for a long term settlement have not progressed.[102] It must be added that the Turkish population controls a segment of the land that is in excess of its population proportion, and the land it controls has a greater concentration of manufacturing. Nevertheless, the income per capita in the Greek section remains higher.

Northern Epirus

When Albania was created and its borders drawn, the largest dispute came from within Greece. This dispute, like others of its kind, was based on the ethnic composition of the resident population. According to the figures compiled by the League of Nations in 1922, Northern Epirus contained 112,329 Orthodox Christians and 113,845 Albanian Moslems.[103] In the present, the size of the Greek population is under dispute: the Greeks claim that there are 400,000 Greeks in Northern Epirus, while Albanians put the figure at 58,000 in 1989.[104] For some years during this century, a segment of what is now Albania was under Greek rule. Indeed, the Protocol of Corfu gave Northern Epirus to Greece. This decision was revoked in 1921 by the Ambassador's Conference, and the region has since been part of Albania. Although this territorial claim is not officially made by the Greek government, there are some public figures that are in favor of the restoration of Greek rule to the region.[105] There is even a U.S. Senate resolution (from 1944) supporting Greece's demands for the return of the region to Greece.[106] The Movement for the Recovery of Vorio Epirus, although not significant in either politics or numbers, is nevertheless a cause for concern in the Tirana government, which views with suspicion its southern neighbor and its potential desires for Albania's dismemberment. It has managed to translate its message to the Greek population, so that the concept of the Albanization of the Greek community in Albania has become an accepted fact.[107] A crisis developed in the summer of 1993 when the Albanian government deported an Orthodox priest from Albania, ostensibly for fomenting nationalism among the Greek population in southern Albania. The move caused outrage in Greece and resulted in the intensification of Operation Broom, an effort to expel illegal Albanian immigrants from Greece. By July 1993, some 30,000 Albanians had been deported, and negotiations are presently proceeding on how to calm the situation. Meanwhile, the Greek population has become hopeful that their condition within Albania has reached the ears of the Greek politicians,[108] and the Albanians fear this new meddling in their internal affairs.

SELF-DETERMINATION ON THE BASIS OF ETHNICITY

Given the lack of congruence between ethnic and political borders and the ethnoterritorial pressures discussed above, one has to consider the ethnic basis for redrawing boundaries and therefore the question of self-determination on the basis of ethnicity. However, while the concept of self-determination of nations was a popular romantic idea in the nineteenth century, soon after its partial application in the real world, it became clear that it was wrought with problems. Two problems stand out. First, if the principle of self-determination is to be applied, then it must be applied to all those who want it. There can be no inconsistency or discrimination. In former Yugoslavia, as pointed out by Zametica, "if the Croats, or Muslims, had the right to pursue self-determination, why not the Serbs? If the Croats and Muslims could secede from Yugoslavia, on what grounds could the Serbs be denied the right to secede from Croatia and Bosnia-Herzegovina?"[109] Moreover, it should not be surprising that, in an interview entitled "All Albanians Must Be Reunited," Albanian foreign minister Serrequi claimed, "We will not hesitate to demand that all Albanians live in their ethnic territories. It is not just to refuse them self-determination."[110] Indeed, if the Serbs in Krajina want self-determination on the basis of their majority status, then why not the Albanians in Kosovo? However, this then implies the proliferation of small nation-states, unviable and marginal. While this issue is discussed at length in chapter 7, suffice it to say here that even Woodrow Wilson, in hindsight, recognized that the consistent application of the right to self-determination to all who requested it would be impractical.[111] Lord Acton also has said that to grant self-determination to all those who request it is "a retrograde step in history."[112]

The second problem has to do with the negative effect that the creation of ethnic states on the basis of the principle of self-determination has on nontitular minorities. Indeed, the pursuit of self-determination in Eastern Europe in the late twentieth century has proved not to be a universally democratic choice. While in a way self-determination seems most democratic, since it allows the individual to determine which country he or she wants to be a part of (even if, in the extreme, the population shrinks to one), in an atmosphere of rabid nationalism, it creates a situation in which democratic rights are conferred by ethnic group, at the cost of nonmembers. According to Etzioni, "The great intolerance breakaway states tend to display toward minority ethnic groups heightens polarization [among groups]."[113] This is true of Estonia, leading to calls of "Estonian apartheid,"[114] as well as of Latvia, Lithuania and Slovakia.[115] Hayden points out that the new successor states of Yugoslavia all have constitutions that are less democratic with respect to minority rights than the last Yugoslav federal constitution before the breakup.[116] For example, the Constitution of Yugoslavia of 1992 claims that the official language is Serbian written in the Cyrillic script (note that, as in the Croatian constitution, the term *Serbo-Croatian* as a language is not used). The Croatian constitution of December 1990 clearly proclaims the rights of the Croatian nation without mention of the 15 percent Serbian population, which was given equality under the previous constitution. In Slovenia, the 1989 amendments to the constitution create, according to Hayden, "a three-tiered set of national privileges: first, the 'sovereign Slovenian nation,' second, the 'autochthonous minorities' [Italians and Hungarians, very small in number], and third (with no constitutionally recognized cultural rights), members of any other national group, who in fact form the largest minority populations in Slovenia."[117] In Macedonia, the constitution of November 1991 is based on the rights of the Macedonian nation and does not grant Albanians, who form up to 40 percent of the population, the same rights as the Macedonians.[118] Thus, nationalist tendencies have become legalized in the constitutions. These undemocratic tendencies have emerged when minorities become majorities and majorities become minorities, since, as Jaszi points out, "the political morals of an oppressed nation change when it comes to power."[119]

Given the above problems of self-determination on the basis of ethnicity, as well as the pressure for boundary changes in the Balkans in 1993, some principles according to which boundary changes will occur must be formulated. This set of principles must then be applied consistently across the regions in which ethnoterritorial pressures are exerted. Indeed, much of the bloodshed in former Yugoslavia might have been avoided if the West had been consistent in its application of principles with which to determine independence, legitimacy of secessionist movements and inviolability of borders.

The inconsistencies and the double standard of western policies toward Yugoslavia can be illustrated by three issues. First, western leaders proclaim to want to protect *the integrity of borders*. This is the principle being applied to regions such as Croatia and Bosnia-Herzegovina. But which borders do they want to protect? The principle has not been applied to the borders of Yugoslavia. By what logic or by the application of which principles does the West decide to protect some subnational borders and not an international border? Indeed, the West did not respect the integrity of Yugoslavia's borders when it recognized Slovenia and Croatia, and was insistent about recognizing the integrity of Croatian borders. This makes no sense, given that there is no reason to believe that internal boundaries of Yugoslav administrative units should be enshrined while international boundaries need not. Furthermore, as long as the Yugoslav federation existed in its political configuration, then there was reason to continue the boundaries as Tito created them. Once that ended, the logic of holding internal boundaries as sacred disappeared. For the sake of comparison, a country such as Switzerland has a constitution that forbids the changing of its international borders. The individual cantons do not have the right to secede. In addition, in the discussion of Quebec's possible secession from Canada, it is openly acknowledged that Quebec will not be allowed to take with it all its present territory. Logically, Croatia should not be allowed to take with it territory whose population chooses not to go, as for example the (Serbian) Krajina region of Croatia. Thousands of Serbs and Croats have died in the battles over this issue.

Second, western leaders support the concept of self-determination as determined with *the use of referenda*. However, they are selective in their choice of *which* referendum is acceptable. The western world was put on the spot with the proposed referendum of the Bosnian Serbs on whether to accept the Vance-Owen peace plan. It was rejected by the West as a sham, despite its obviously democratic aspects.[120] While the referendum of the Slovenian population on its secession was acceptable as the basis for international recognition, the referendum of the Serbs of Krajina was not. Furthermore, the western leaders have no clear policy on *who* is to be allowed to vote in a referendum to make it acceptable. In all likelihood, if all citizens of Yugoslavia participated in a vote on Slovenian secession in 1990, the outcome would not have been in favor of secession. For the purpose of comparison, it should be recalled that in 1979 the entire French population voted in the referendum for the independence of New Caledonia, not only the islanders. Moreover, the constitution of Spain, a country that has recognized the seceding republics of Yugoslavia, explicitly prohibits a referendum on secession of any of its regions.

Third, western leaders seem unable to determine the conditions under which to grant *international recognition* to newly seceded regions. The EC hastily drew up a shopping list of conditions, including the protection of rights for minorities, and proceeded to recognize Croatia, Slovenia and Bosnia-Herzegovina, despite France's short-lived objections that Croatia did not fulfill the minority-protection criterion. At the same time, the western powers have yet to fully recognize Macedonia, which with Slovenia is the most deserving of recognition if judged by the minority issue. Western hesitation in the Macedonian case is due to Greece's rejection of Tito's creation of a "Macedonian identity" within Yugoslavia. However, other creations of Tito's are not rejected, such as his redrawing of internal boundaries. This can only lead to further deterioration of the situation. According to Simmie and Dekleva, "The [European] Community cannot contribute to destabilization nor create potential difficulties for its own members by interfering in the internal affairs of extra-Community nations. Such interference would include the recognition of parts of existing states as separate nations."[121] Dyker adds, "Of one thing we can be sure. Any western attempt to take sides in the political dramas now unfolding would almost certainly have a perverse effect."[122] Indeed, even Bosnian president Alia Izetbegovic was not in favor of early recognition of any Yugoslav republics.[123]

The action of the western leaders pertaining to the Yugoslav conflict, steeped as it has been in double standard and inconsistencies, has directly contributed to the escalation of the war in Yugoslavia. While this will be discussed at length in chapter 7, suffice it to say that the new world order broke in Yugoslavia. In the beginning, German support of Croatia was too strong, too reminiscent of World War II and thus clearly counterproductive to a peaceful solution involving the Serbs. Later, EC and U.S. recognition of Croatia and Bosnia-Herzegovina reduced the options available to Serb residents of those areas who favored remaining part of Yugoslavia, making the military option seem like the only salvation for some. Sanctions and international isolation further intensified Serbian resolve as Serbian fighters realized they have less and less to lose. Thus, western action aided the demise of the very idea of Yugoslavia. The western leaders needed to intervene, with a consistent application of principles, at the time of the emergence of nationalist leaders in Slovenia, Serbia and Croatia. They might have foreseen that nationalist states based on ethnic purity were not a viable means to peace and might have used their power to help promote the idea of a larger unions characterized by 'unity in diversity,' rather than to support nationalists who fanned intolerance and hatred in peoples heretofore largely peacefully coexisting.

CONCLUSIONS

It is clear now, with so many ethnoterritorial movements raging across the Balkans, that mere border changes will not and cannot satisfy everyone. Indeed, what Roucek wrote in 1939 is still true: "Boundary changes have never settled the problems of all groups involved. Bulgaria mourns Dobrodja and Macedonia, Greece weeps over Epirus; Hungarians have been proclaiming vociferously their *nem nem soha* (no no never) over the lost Transylvanian districts."[124]

It is also clear that principles for implementing boundary changes have to be chosen on the basis of which ethnoterritorial movements will be recognized. If the principle of ethnic majority (demography) is adhered to, then Croatia stands to lose parts of Krajina and parts of Istria; Montenegro stands to lose parts of Muslim-populated territory to the west; Serbia stands to lose some districts in Vojvodina and the Sandzak, as well as Kosovo; Bosnia must be partitioned into three parts, with the Serbs getting the majority of the rural areas while Muslims get the majority of the urban regions and parts of Herzegovina go to the Croats. Macedonia would also be split, as eastern regions are inhabited by Albanian Muslims and western areas by Bulgarians. And this is only Yugoslavia—if the principle were applied to the entire Balkans, a similar pattern would emerge. A second principle is possible, namely that of historical rights. The first question this would raise is how far back must one go? Invocations of the Illyrians and the Bogumils raise questions that can easily be disputed. Indeed, each region has its period that it wants to glorify.

The process of border changes in the Balkans has been going on for centuries, and no single power will be satisfied without the restoration of the boundaries it enjoyed at its peak, which would amount to enlarging the area of the Balkans fivefold. Therefore, for the sake of peace, a supernational government that gives all ethnic groups a sense that they are indeed ruling along those borders should be devised—in other words, some form of multinational state with power sharing. On the verge of the twenty-first century, the Balkans should be able to come up with political and social systems that can accommodate differences in ethnicity and come up with viable political and economic solutions to border problems.

3

Manifestations of the Balkan Economic Crises

"The challenge of the epoch is to mesh the centripetal forces of economics with the centrifugal ones of politics."

Fred Bergsten[1]

The Balkan economic turmoil of the early 1990s manifests itself in a wide variety of economic problems, including drops in economic growth and industrial production, unemployment, inflation, indebtedness and the proliferation of underground economies. According to the evidence pertaining to these indicators of economic health, there is an economic crisis in the Balkans that is yet to bottom out (with the possible exception of Slovenia). The prevailing economic turmoil has its roots in a variety of internal and external sources: While the transition to capitalism that permeates the formerly socialist states is certainly a factor in that turmoil, as is the disruptive effect of the civil war in Yugoslavia, these by no means mark the onset of the problem, but rather constitute the consequences that in turn perpetuate and deepen the crisis.

There is evidence that the Balkan economies have been in a state of crisis for years. Indeed, to speak of an economic crisis today in this part of the world is almost meaningless: Yugoslavia has reportedly been in "perpetual crisis" since the 1980s;[2] Greece has been expecting a "takeoff that never was."[3] Albania was rumored to have "collapsed" numerous times in its postwar history, while many Romanians would claim that they are in perpetual collapse.[4] However, the term *crisis* connotes urgency and severity of difficulties. It also connotes a sudden onset, and for that reason it is again appropriate to use the term in association with the deepening and broadening economic turmoil of the late 1980s. There has, in fact, been an increase in urgency since 1989. That is the year associated with the end of communism in Eastern Europe, the demise of the Soviet lifeline to the Balkan economies, and the end of the cold war, with its political ramifications for the Balkan political economies. These changes in the external environment, together with internal ones, put pressures on the Balkan states for dramatically restructure their economies. Furthermore, by 1990, the Yugoslav federation began to crumble, and the economic effects of that disintegration, with its concomitant war and sanctions, began to have a multiplier effect throughout the Balkan economies, further aggravating each individual crisis.

The present economic turmoil cuts across levels of development as well as economic systems, affecting Slovenia as well as Kosovo, Greece as well as Romania. This commonality of economic malaise further strengthens the arguments for studying the Balkan states in unison. The lack of this common element in the period 1945 to 1989 was in part responsible for the paucity of economic studies pertaining to the entire region. Indeed, the economic literature tended to focus on Balkan countries individually or Eastern Europe as a whole. Greece's status as a capitalist economy segregated it from studies of the socialist states, and its membership in the European Community warranted its inclusion in the EEC literature. The rare comparative studies of communist and capitalist countries of the Balkans include the historical analysis by Lampe and Jackson, an overview by Gianaris, Hoffman's study of spatial development and, more recently, an economic focus by Sjoberg and Wyzan.[5]

MANIFESTATIONS OF THE CRISIS

Economic Growth and Economic Development

The Balkan states, for centuries in the domain of Ottoman Turkey and Hapsburg Austria, were still agricultural lands at the turn of the twentieth century. While parts of former Yugoslavia experienced industrialization in the nineteenth century, agriculture remained the principal source of employment and state income. Structural imbalances began surfacing as the population grew and demand outstripped the available resources of the region. This imbalance did not change until World War II and the new policies that followed it. In all Balkan states except Greece, the postwar period introduced a new economic system. Socialist economic relations and organization tended to be focused toward rapid industrialization of traditionally agrarian economies. For example, in Romania and Bulgaria, policies of industrialization resulted in an increase in the share of industry: between 1950 and 1970, the industrial share of income in Bulgaria rose from 37 percent to 49 percent, while that of Romania rose from 44 percent to 60 percent.[6] The structural transformation often associated with economic development oddly did not take place in Greece as it did in its Balkan neighbors. Despite efforts by various post-World War II governments, including tax shelters, export subsidies for industries, and five-year plans, industrialization simply did not take off. Indeed, while in 1960, manufacturing and construction accounted for 25.2 percent of the GNP, in 1990 this dropped to 22.8 percent. At the same time, services rose from 30.7 percent to 44.9 percent, while agriculture dropped from 25.9 percent to 14.8 percent.[7] The structural transformation during this period stands in sharp contrast to that of the 1930s, when, according to Mazower, Greek industrial growth rates were surpassed only by Japan and the Soviet Union.[8]

Rates of economic growth have varied greatly in the Balkan states. While Bulgaria experienced high rates of growth during the early 1970s, by the end of the decade growth slowed down, reaching -1.8 percent in 1985. Yugoslavia and Greece enjoyed stellar economic growth: the former during the 1970s, and the latter in the 1960s, when it witnessed the highest rates of growth in the western world, second only to Japan. Indeed, the Greek economy seems to have experienced dramatic swings in economic performance: a dramatic crisis and default in the 1930s was followed by an even more dramatic recovery in the 1960s.

Since the late 1980s, with the demise of communism, there has been a further drop in the growth rates of all Balkan states. In Albania, industrial production fell to only 35 percent of its level in 1990.[9] Only one-third of its 300 largest enterprises are now operating. Its total growth rate for 1991, as evident in table 3.1, is -21.1 percent. In Romania, the overall growth rate, while negative, was higher than industrial production, which in October 1992 was down 23.5 percent from the level of September 1991 (although it was up 5.5 percent from August).[10] Bulgaria experienced a dramatic drop in growth over the period 1988 to 1991, as did other Balkan states. Of all the former Yugoslav states, Slovenia has experienced the smallest drop in its growth rate, namely—11 percent in 1991, and, according to official statistics, this has improved to -6.5 percent in 1992.[11] If one were to observe GDP per head in 1992 based on purchasing power parity, then Slovenia leads; followed by the Czech Republic, Hungary, Poland, Croatia, Serbia and Albania, in that order.[12] Macedonia experienced a drop of 30 percent in industrial output in two years (1990-92).[13] Croatia experienced a drop in total production of 30 percent in 1992, while in the high-tech industries the drop was 70 percent, and iron and steel production came to a virtual halt.[14] Economic growth in Serbia and Montenegro, together with the rest of former Yugoslavia, shows a decline, as national income dropped 15 percent in 1991. This was mostly industrial decline (21 percent) and a decrease in tourism (61 percent), since agricultural output increased about 8 percent over 1990.[15] In 1992, the drop in GDP was 25 percent, while industrial output fell by 22 percent (some industries, such as vehicle manufacturing and nonferrous metallurgy, declined 50 percent over 1991).[16] According to table 3.1, Greece is the only Balkan state that has achieved positive rates of growth in 1991, albeit following a year of negative growth. The latest figures for 1992 indicate a growth rate for Greece of 1 percent.[17]

These dismal growth statistics, coupled with information on inflation and employment (see below), lead to the inevitable conclusion that the standard of living of the Balkan populations has fallen in the course of the early 1990s. A few examples clearly indicate this. In Romania, it is estimated that in 1992, 42 percent of the country's families were living at or below subsistence level.[18] In Macedonia, the average monthly wage has fallen to the equivalent of DM 100 (in mid-1993), and to DM 30 and DM 20 in Serbia and Montenegro, respectively.[19] In Croatia, real household income is currently less than one-third of its 1990 value.[20]

However, this assessment of the standard of living does not take into consideration the wealth created by the underground economy. This illegal economy has been expanding, especially after the demise of communism. While governments are concerned with it because it hurts them (by decreasing tax revenue, for example), it simultaneously helps them by alleviating the pressure to supply consumers with goods and services. Thus, the underground economy has become crucial across the Balkans since an ever-increasing quantity of output is produced by this type of economy, especially in the consumer goods market. This economy includes illegal activities, especially sanction busting with the new Yugoslavia, as well as barter, since there is a cash shortage to pay for items. Minic, who recently described the black-market economy in Serbia, estimates that in 1992, the hidden economy's share of total economic activity was 29.4 percent.[21] Popovic estimates that the value of transactions in the underground economy exceeds that of the official economy.[22] In Macedonia, the black market is thought to account for up to 30 percent of the GDP, in large part due to sanction busting and tax evasion.[23] In Greece, the underground economy is also estimated at around 30 percent of GDP.[24] This makes it the richest underground economy

Table 3.1
Economic Indicators of the Balkan States

State	Growth (percent [year]) (a)	unemployment (percent)	inflation (percent [date])
Albania	5.8 (1989) -21.1 (1991)	NA	6 (June 1992) (g)
Bosnia-Herzegovina	NA	NA	86 monthly (April 1992) (b)
Bulgaria	5.8 (1988) -25.0 (1991)	13.5 (c) (1992)	100 (1992) (c)
Croatia	3.4 (1988) 30 (1991)	20 (f)	40 monthly Dec. 1992) (f)
Greece	-0.1 (1990) 1.8 (1991) (h)	8.1 (1992) (h)	18 (1991-mid 1992) (h)
Macedonia	-18 (Jan 1992) (d)	20 (d)	70 monthly (May 1992) (b)
Romania	0.8 (1987) -8.0 (1991)	8.7 (c) (1992)	250 (1992) (c)
Slovenia	-0.2 (1988) -11.0 (1991)	12-13 (i)	92.7 (1992) 3000 (1991) (i)
New Yugoslavia	-0.5 (1989) -15 (1991)	19 (1991)	235 (1991) (e) 102 monthly (May 1992) (b)

Sources: (a) Political Risk Services, IBC USA Publications, *East Europe and the Republics,* July 1992.

(b) RFE/RL Research Report, January 15, 1992, 34. For lack of disaggregated data, inflation for Bosnia-Herzegovina is derived from inflation in the entire dinar zone, of which it was a part at the time.

(c) RFE/RL Research Bulletin, vol. 10, no. 1 (January 5, 1993): 1.

(d) RFE/RL Research Report, vol. 1, no. 25 (June 19, 1992): 37.

(e) *PlanEcon Report,* vol. 8, nos. 14-15 (April 14, 1992): 2.

(f) *Euromoney* supplement (May 1992): 6.

(g) *Transition,* vol.3, no. 7 (July 1992): 17.

(h) OECD, *Economic Survey of Greece* (Paris: OECD, 1992).

(i) Data released by the Slovenian Embassy in Washington (February 1993) from the Ministry for Economic Relations and Development.

of the EC countries. Thus, anyone considering the statistics on the standard of living, as well as estimates of GNP per capita and economic growth, must recognize that large portions of the Balkan economies do not pass through the formal markets and therefore are not reflected in these numbers.

Unemployment

The changes in internal and external conditions associated with the early 1990s have resulted in dramatic increases in the number of people out of work across the Balkans. While the proximate reasons for this unemployment have varied across the states, the manifestations are similar: large numbers of people abruptly unemployed, underemployment, paid leaves and an unsatisfactory social net to tend to the unemployed. In the new Yugoslavia, unemployment at the end of 1991 reached two million people, approximately 19 percent of the population. By the beginning of 1993, this number had risen to 25 percent, resulting in strikes of workers demanding compensation when laid off.[25] In February 1993, unemployment approached 40 percent.[26] And by August 1993, Andrejevich reported that more than half of the workforce was either unemployed or on extended leave of absence.[27] In Montenegro in spring 1993, there were 68,000 unemployed workers, 70,000 pensioners and 65,000 refugees out of a total working population of 400,000.[28] A similar trend, although less dramatic, is also evident in Croatia, where employment fell 30 percent in 1992.[29] In Slovenia, the unemployment rate was 15 percent in mid-1992.[30] It was at 14 percent in early 1993, and is expected to rise, as half of Slovenia's 23,300 companies, which employ more than 486,000 workers, are on the brink of collapse.[31] In Albania, only one-third of the country's 300 largest enterprises are operating, and unemployment in industry has reached 50 percent.[32] In Romania, in October 1992, some 8 percent of the labor force was unemployed, amounting to 860,000 people.[33] In Bulgaria, unemployment had reached 13.3 percent in September 1992,[34] and it reached 17 percent as of June 1993.[35] In Macedonia, there were 16,000 job losses in April 1993, when 1,300 companies filed for bankruptcy.[36]

This rise of unemployment in postcommunist states is due largely to measures associated with the transition from socialism to capitalism (see chapter 4) as well as the stabilization policies, such as the closure of inefficient, heavily indebted enterprises. Not only does this leave people without work, but it also often leaves them without a safety net. While during communism, unemployment compensation was one social benefit, under the stringent conditions associated with the early 1990s, such benefits can no longer be taken for granted. The Romanian example is a case in point. There, less than half of the unemployed receive unemployment benefits. Nevertheless, this unemployment contributes to the fact that 43 percent of the population is considered to be living below the poverty line.[37] While Romania's trade unions have been very active in mid-1993, partially in a quest for employment and unemployment benefits and to address other hardships associated with the transition to capitalism, they have yet to prove effective in dealing with a government that itself is financially constrained. The Romanian example is repeated throughout the Balkans and unemployment has thus been coined the "number one social problem" of postcommunist countries. It is a problem in Greece also, despite the relatively low rate of unemployment. At least some of this unemployment can be attributed to the austerity measures adopted by the Mitsotakis government. These measures, discussed in greater detail in chapter 4, were similar to those adopted by formerly socialist states in their transition to capitalism: namely, the trimming of the public sector and the weeding out of inefficient firms. In addition, the government was to alter a wide variety of measures affecting workers, including pensions, retirement age and benefits. The labor force in Greece voiced its opinion of liberalization and austerity measures that were adopted in 1992 by paralyzing the country with strikes in mid-1993.[38]

Inflation

All Balkan states, with the exception of Greece, are experiencing high rates of inflation. In Romania, inflation was 19 percent in early 1992.[39] The total yearly inflation rate in Romania in 1992 was 250 percent, compared with 200 percent in 1991.[40] From October 1990 until October 1992, food prices increased by 1,074 percent.[41] In Bulgaria, inflation in 1992 reached 100 percent, down from 500 percent the preceding year.[42] In Albania, inflation is currently running at an annual rate of 220 percent.[43] In Croatia, the monthly rate at the end of 1992 was 30 percent.[44] The Croatian dinar, introduced in December 1991, is expected to be replaced by the Croatian kuna by the end of 1993,[45] as part of a plan to stabilize prices. In Macedonia, during the time that the new country was part of the dinar zone, inflation reached 86 percent monthly (April 1992). However, Macedonia left the dinar zone, introduced its currency (the denar) and began a successful anti-inflationary program; inflation dropped to 70 percent in May 1992 and 17 percent in June 1992.[46] The further measure of devaluation of the denar by 67 percent in October 1992, brought the monthly inflation rate to between 20 percent and 30 percent.[47] Slovenia has experienced a low inflation rate and like Macedonia, it experienced a drop after leaving the dinar zone in October 1991.[48] However, it did not need to devalue its currency, the tolar, in the first nine months of its existence. The inflation rate in Greece is 16 percent in mid-1993. While this is not high relative to the Balkan states, it is the highest of the EC, raising disconcerting questions about Greece's membership in the Community. Like other Balkan states, Greece not long ago had one of the lowest inflation rates in Europe, namely during the 1950s and 1960s. Measures taken by the government lead to the expectation that Greece's inflation rate would fall to below 10 percent by the end of 1993.[49]

The most dramatic case of price rises has been experienced by Yugoslavia. Indeed, the recent experience of Yugoslavia surpasses all textbook examples of galloping inflation, even that of the Weimar Republic in 1923. The inflation figures released by the government in August 1993 indicate that the July 1993 rate was 1,880 percent, amounting to an annual rate of approximately 1.7 billion percent.[50] In March 1993, the annual rate of inflation was only 25,000 percent.[51] The prices of some 2,000 items rose on March 1, while electricity prices rose 145 percent. This galloping inflation led the national bank to issue a one billion dinar bank note, which is worth only $3 at the current exchange rate.[52]

Trade

All Balkan states, with the exception of Albania, had interdependent economies. Trade flows have been disrupted in the 1990s, and the decrease in trade revenue has had a negative effect on the Balkan states. States vary in the degree to which they have been affected. Indeed, while some states have trade surpluses (notably Romania, with a trade surplus of $68.2 million in October 1992, and Slovenia, with a $161 million trade surplus in March 1992),[53] other Balkan states experienced trade deficits (notably Yugoslavia, excluding Slovenia, with a $1 billion merchandise trade deficit in September 1992).[54] With respect to the volume of trade, there has been an overall decrease across the region. For example, in the former Yugoslavia, imports fell by 27 percent, while exports fell by 21 percent during 1991.[55] But perhaps most interesting is the change in trading partners: Romania has turned to Serbia in a forming of economic ties that would have grown further in the absence of sanctions; Bulgaria has turned to Turkey; Macedonia is developing economic ties with Albania and Bulgaria to offset the loss of Greek and Serbian borders as well as a major contract with Turkey;[56] and Albania is cultivating Italian and Turkish linkages in an efforts to open its economy.

What caused the changes in Balkan trade? First, the breakup of Yugoslavia has meant the loss of markets for imports and exports of the various former republics. The exact magnitude of the impact associated with this breakup is hard to assess: there has been evidence that all economies, ranging from Slovenia's to Macedonia's,[57] have suffered from the loss of markets. There is not even clear evidence of the importance of markets before the breakup: studies were conducted by Bicanic, Bookman, Ding and Ocic. All point out that there was a disintegration of markets, but they varied in their assessments. According to my study, 30 percent of Slovenia's trade was with the external markets.[58] According to a study by Grubisic, Croatia sold 68.6 percent of its goods within Croatia in 1987, 19.1 percent to the Yugoslav markets, and 10.8 percent in the international markets.[59] According to Bertoncelj-Popit, for Slovenia, Croatia was the largest market, accounting for 60 percent of its trade with Yugoslavia.[60] Perhaps most complete are the data compiled by Milojcic, which indicate that over 60 percent of trade was among the republics.[61]

Second, the sanctions imposed on Yugoslavia are dramatically altering inter-Balkan trade relations. Yugoslav trading ties have dwindled to those that can persist in the semilegal atmosphere of sanction busting. In practice, this has meant that Greece, and to a lesser extent Macedonia and Romania, have continued, albeit at a diminished rate, to trade with Serbia and Montenegro. All Balkan countries are negatively affected by the sanctions since transportation routes through Yugoslavia have been severed. Other roads—for example, from Greece to the north—lead through the war zone or are excruciatingly circuitous. One response to the problem is the Tirana-Istanbul highway that is in the planning stage. It would be a substitute for the traditional transportation routes, and would link Albania, Macedonia, Bulgaria and Turkey in a direct trading route. This would facilitate new trade patterns in all states, especially Macedonia, which has been doubly squeezed by sanctions to the north and the Greek embargo to the south.[62]

Third, with the exception of Greece, the Balkan trade pattern that has predominated in the past two decades has been one that centered on the CMEA. This intensification of CMEA trade was due, in part, to the oil shocks of the 1970s, followed by the stagnating western economies and the debt incurred by the states. Before dissolving, the CMEA introduced regulation pertaining to the use of convertible currencies instead of unconvertible Eastern Bloc currencies in all its trade relations. Moreover, credits were discontinued, causing strain on cash-short states. Bulgaria and Romania were the two states most dependent on trade with the CMEA, and thus its dissolution left those states in the most vulnerable positions. Indeed, in 1988, 58 percent of Bulgaria's exports went to the Soviet Union, and 53 percent of its imports came from there.[63] Even for Yugoslavia, the Soviet Union became the principal trading partner in the 1970s. Clearly, the new regulation put into effect on January 1, 1991, that all trade between the Soviet Union and its CMEA partners would be conducted for hard currency had a detrimental effect on the Balkan states insofar as the volume of trade decreased. Some lost trade has simply been replaced by trade with the CIS states. At the same time, it has become clear to all postcommunist Balkan states that western markets will not replace CMEA markets, at least not in the short run. Indeed, breaking into the highly competitive western markets is proving harder than previously imagined, as clear from the comments by Smith: "Balkan socialist states are chasing a moving and even accelerating target in terms of their competitive position in the world economy."[64] To rectify this trend, there have been some minor bilateral changes in the trade patterns of the Balkan countries as a result of these changing conditions. For example, Bulgaria signed an accord with the European Free Trade Association for the abolishment of trade barriers in industrial goods and processed agricultural products, effective July 1993. Given that Bulgaria's trade with the EFTA countries in 1992 was $102 million in exports and $203 million in imports, such an agreement is bound to have a large effect on the Bulgarian economy.[65] Romania has already signed a similar accord. Another example of these new trading ties is the one forged between Albania and Turkey. It consists of a 15-year economic development program to which Turkey has committed itself, which includes the renovation of its ports, development of tourism and financial institutions and military cooperation.

Foreign Investment

Since 1989, foreign investment in the Balkans has sharply decreased, aggravating the economic crisis. Despite the transition to capitalism that has been advertised to potential investors, and the concomitant relaxation of laws pertaining to foreign companies, the net effect of the investment into the region has been negative. Investment simply is not flowing into the former Eastern Bloc nations: while Mexico, Argentina and Chile attracted $7.8 billion in direct foreign investment in 1991, the Czech Republic, Poland and Hungary, the most successful of the former states, attracted merely $2.2 billion.[66] Despite the fact that German foreign investment is up to its pre-World War II levels, as measured by PlanEcon, the quantities have been disappointing in the Balkan states, which are at a disadvantage relative to the more reformist north-central Europe. Even Slovenia, which was expected to be the exception, is shown by Zapp to be very disappointed at the lack of sufficient foreign investment. In Bulgaria, in 1992, there was $41.5 million of direct foreign investment, and the number of joint ventures with foreign firms was 837 as of May 1993.[67] In Romania, the numbers are significantly larger: foreign investment was valued at $680 million as of June 1993, and the number of joint ventures by that date were 25,484.[68]

Why is there not more public and private foreign investment in the Balkans? First, the war in former Yugoslavia has caused turmoil and scared away investors in a radius wider than the actual warring regions. Second, the sanctions in new Yugoslavia preclude investment on Yugoslav territory and prevent investment in surrounding regions that are adversely affected by those sanctions. Clearly, the economic deprivation caused in Romania by the sanctions will translate into an uninviting economic climate for investment. Third, states such as Macedonia could not apply for international aid and funding since they were not recognized as sovereign entities. In order to offset this debilitating limitation, Macedonia received a loan from the Soros Foundation of $25 million to purchase fuel and supplies for winter 1993.[69] Fourth, some investing states are saturated and overwhelmed with both demands at home (e.g., Germany after unification), and demands from other states (e.g., the rest of Eastern Europe). This means that the Balkan states must compete with numerous other countries for foreign investment. Fifth, the Balkan states are not progressing as quickly and thoroughly as might have been hoped with the reforms toward a market economy and thus still represent potential quagmires for foreign investors. All of these factors imply that the Balkans have lost their appeal in the world of international finance. They are seen as a troublesome region that cannot settle its petty problems and is strategically no longer of consequence. Speaking of the former Yugoslavia, Cviic says that there is an "economic Yugoweariness in 1990" as the West is no longer willing to infuse the country with money.[70] That applies to the rest of the Balkans also, with the exception of Greece and, to a smaller degree, Albania. Greece is the recipient of a constant inflow of money from the European Community, which it joined in 1981, while Albania is the recipient of foreign aid rather than investment.[71]

Tourism

Tourism is an important source of income for many Balkan states, and its disruption in the early 1990s has had a major effect on Balkan economies. Tourism is a principal source of foreign currency and a major contributor to state income. Moreover, its development has a multiplier effect on regional economies, as it tends to build a supporting infrastructure, such as electrification, transportation, housing, et cetera. Greece and Croatia are especially dependent on tourism as a source of income and foreign currency. In 1989, Greece acquired an inflow of $591 million from tourism.[72] According to official statistics, tourism in Croatia accounts for some 12 percent of GDP and employs between 10 and 12 percent of the population.[73] Given the configuration of transportation routes in the Balkans, each region enjoyed a spillover effect of each other's tourism. Indeed, Slovenia profited from tourists bound for the Dalmatian coast or Greece; Serbia profited by providing overland routes for *gastarbeiters* (temporary guest workers in Germany) on their way to Turkey; Macedonia represented a stopover for Greece; and Romania and Bulgaria were attractive to travelers from the Eastern Bloc whose unconvertible currencies failed to suffice for vacation further west or south.

The Yugoslav wars of succession, and to a lesser degree the recession in Western Europe, have had a varied effect on tourism across the Balkans. Croatia suffered the most: during the civil war, the number of overnight stays fell to 54 percent of the 1990 level, and earned the economy $298 million.[74] While many of these tourists were from Western Europe, the importance of domestic tourists should not be underestimated, especially in former Yugoslavia, where in 1987, 60 percent of the tourists were domestic.[75] Greece received one-third fewer tourists in the summer of 1993 than the previous summer. The principal travel route to Greece had been by automobile through Yugoslavia; this route is presently inaccessible and dangerous. According to Yannis Anglos, "No one can set off on his holidays through Yugoslavia and know that he will remain unaffected by the situation all along his route."[76] On the other hand, there has been a significant increase in tourism in Bulgaria. Indeed, during the first nine months of 1993, there was a 42 percent increase in total visitors to Bulgaria over the same period in 1992, and a 48 percent increase in visitors from western countries.[77] Most of these visitors are from Germany and Britain, the two principal sources of tourists in both Yugoslavia and Greece.

Defense Budget

Given a stagnant economy, an increase in the defense budget implies less income to be allocated for other competing ends. Under conditions of decreasing income, as has been the case in the Balkans in the 1990s, then an increase in defense spending jeopardizes even more the health of other sectors of the economy. The necessity of focusing finances and energy on the military sector has contributed to the crisis in Balkan states. Clearly, the proportion of the budget allocated to defense is the largest in those Balkan states most affected by the Yugoslav wars. In mid-1993, 75 percent of the federal budget in Yugoslavia was earmarked for the Yugoslav Army. The extent of aid supplied to the Serbs in Krajina and in Bosnia is not clear, but it is not zero. Furthermore, threats from the United States of military intervention, hinting even at incursions on Serbian territory, have not helped the argument in favor of decreasing the military spending. In Croatia, military expenditures constitute at least 30 percent of the state budget.[78] Not only has Croatia been at war with the Serbs in Krajina in 1991, but it is also actively supporting the Bosnian Croats fighting in Bosnia. Of all the successor states, Macedonia is the least concerned with defense at this time. While it has historically been in a perilous position, at this time it has no army and no allocation for defense in its budget.

However, even in peace, Balkan countries tend to allocate large portions of their budgets to defense. The fact that they tend to be at war at some point in any given individual's lifetime points to the necessity of standing armies. This is best exemplified by Greece, which allocates 6.2 percent of its annual income to defense, three times as much as other members of the EC.[79] This military preparedness is due to fear of Turkey, with whom it is locked in perpetual hostility.

CONCLUSION

The macroeconomic picture painted above indicates some dimensions of the crises in the Balkan states. To this must be added the complicated monetary picture, as well as microeconomic problems, including the shortage of investment resources, the lack of competitiveness, the lack of new technology and the lack of know-how. Together, these factors paint a sad picture, with the possible exception of Slovenia, where the crisis may have bottomed out. This positive assessment has led *The Economist* to call Slovenia "the one that got away."[80] According to Mencinger, a Slovenian economist, the economy is bottoming out and will stabilize in the next year or two, as long as the recession in western markets does not intervene too much with Slovenian exports and tourism picks up.[81] Slovenia perhaps achieved this measure of success relative to other regions because it was at an advantage to begin with. It enjoyed a higher income per capita than most regions, and it enjoyed economic and political support from its immediate western neighbors. In addition, it is geographically isolated from other Yugoslav republics and was not involved in the war. Contrary to the Slovenian case, there has not been a bottoming out in the rest of former Yugoslavia, as neither the Macedonian, Croatian nor new Yugoslav economy shows any signs of picking up. While the Macedonian economy has proved more resilient than expected, insofar as it was able to withstand the loss of the Yugoslav markets and the Greek embargo, it nevertheless received a dismal report from the World Bank Economic Mission in spring 1993. The bottoming out of the Serbian economy is certainly not yet in sight. Not only is the region plagued by the consequences of war, sanctions and transition, but it is also suffering from structural limitations: according to Adamovic, Serbia is in the "zone of frustration," due to its size and its limited natural and human resources. He writes, "It is not small enough that the need for a specific specialization would spring to mind . . . [nor] is it big enough that for a broader range of programmes it could be self-reliant."[82] Together, these factors have hindered efforts at resolving the economic crisis. Nor do any other Balkan economies show signs of bottoming out. The experience of Romania is a case in point: while mid-1992 seems to have been a turning point with respect to industrial production, 1992 was also the third consecutive year in which both inflation and unemployment increased, and the standard of living, as indicated by growth rates, declined.

In multiethnic societies, dire economic conditions such as those portrayed above cannot but evoke a sense among the populations that their particular ethnic group is disadvantaged due to the advantages of another group. The sense that well-being is a zero-sum game in which one group gains at the expense of another is exacerbated during hard economic times, indicating that it is not a coincidence that interethnic animosities go hand in hand with economic deprivation. This is most clearly exemplified in the case of former Yugoslavia. There, economic problems are at the root of the present quagmire, and they exacerbate interethnic relations. Prior to the economic crisis of the 1980s, constituent ethnic groups interacted with little friction, and nationalist elements were fringe elements. However, at a time when economic stagnation precluded the growth to which regions had gotten accustomed, perceptions of economic injustice developed along regional lines, which coincided with ethnic lines. Leaders that fostered and harnessed feelings of nationalist deprivation emerged and drew attention away from the economic issues. The issue of Macedonia and its role in the Greek political forums presently is another example of such diversions in the face of economic stagnation and mismanagement.

The current economic quagmire in the Balkans is reminiscent of the economic problems plaguing numerous Third World countries. Indeed, the macro and micro obstacles to economic development are not dissimilar to those that have proved intractable and long-term in developing countries. While this similarity, after decades of socialist development, may be disheartening to those who viewed socialism as an alternative to the ravages of pure capitalism, it need not be disheartening with respect to its future outlook. Indeed, the Balkan states, with the possible exception of Albania, have numerous advantages over most developing countries. First, they have a more skilled and educated labor force that is capable of easier adaptability and a higher overall level of modern production. Second, they have a better-developed infrastructure, such as banking, transportation and communications. Third, they do not face the population pressures faced by most of the developing countries. Indeed, throughout the Balkans, with the exception of the Albanian population in Kosovo, the birth rates have been low and the concern is more one of too little growth rather than too much. Fourth, they are part of Europe, and Western Europe has an interest in the stability and development of the region. Therefore, they are more likely to receive grants and loans than some Third World countries that would have to rely on regional powers, such as Japan in Asia or the United States in South America. The Balkan states do face certain disadvantages: for example, the laborious task of unraveling the socialist infrastructure before building a new one, as well as the breaking down of barriers against a conversion to capitalism. However, these cannot be said to outweigh the advantages, and the outlook for the economic reconstruction and development of the Balkans is much brighter than in developing countries.

4

Former Yugoslavia: Transition to Capitalism and War Economics

"Yugoslavia has now passed the turning point. It is deep in an irreversible but smooth transformation or 'soft-landing' into market economy and multiparty democracy."

Zivko Pregl, Vice-President of the Federal Executive Council of Yugoslavia, July 1990[1]

"[A] reform program is like riding a bicycle: you have to keep moving rapidly, or else you fall off."

Larry Summers[2]

"The ugly underside to the transition to democracy [is] war, political fragmentation and economic despair."

Charles Gati[3]

The entire (formerly) communist world is reeling under the pressure for marketization. Even countries that have withstood change in the initial years of the transition to capitalism, such as China and Cuba, are presently reconsidering their positions. Since the late 1980s, some basic divergences have been emerging among formerly communist countries with respect to the speed and depth of reforms. The Balkan states, with the possible exception of Slovenia, have all tended to lag behind in the transition process. States are also divided into those that have the option of pursuing transition reforms and those that have of necessity concentrated their energies elsewhere, such as on war. This applies especially to Croatia, the new Yugoslavia, and

Bosnia-Herzegovina, although peripheral Slovenia and Macedonia have also succumbed to the ripple effects of war. These effects have even permeated neighboring states such as Romania, Albania and Bulgaria. Thus in former Yugoslavia three developments are at odds with each other and are having mutually exclusive effects on the economy: first, the transition from socialism to capitalism; second, the movement to a sovereign economy, and third, the transformation of a civil society to a war economy. Each of these developments is accompanied by a set of demands on the economy and population, policy measures and costs. Their occurrence individually would cause a strain on any society; their occurrence simultaneously is devastating. Indeed, in scope and breadth, the developments in the former Yugoslav republics surpass even the revolutionary changes in Tibet, when "as never before in history, a nation stood up to oppose both feudalism and communism . . . it was a double revolution . . . the Hungarian and French revolution combined."[4]

This chapter focuses only on two of these developments, namely the transition to capitalism and the transformation into a war economy. The changes associated with movement from federal union to sovereignty have been discussed by the author in *The Economics of Secession.*[5]

THE TRANSITION PROCESS

In the late 1980s, economists bemoaned the lack of theory describing the transition from socialism to capitalism. Indeed, while theory describing the process of moving from capitalism to socialism existed, scholars were caught unprepared for the reverse real-world events of the late 1980s. However, the early 1990s witnessed the proliferation of writings on the extent, pace and feasibility of reforms aimed at introducing capitalism to formerly socialist states and thereby incorporating them into the world economy. The end of the cold war and the professed desires of former Soviet Bloc states to embrace capitalism signified to many the victory of democracy and capitalism and thus the end of the need to study this region separately.[6] Yet the rapidly changing events associated with the rise of nationalism, the economic hardship and political realignments following the cold war all point out that, to the contrary, the attention of research must once again be focused on these changing states and their emerging systems.

From this proliferation of literature on the process of economic transformation to capitalism, several basic tenets may be distilled. The process of transition involves the following three categories of change: macroeconomic stabilization, price reform and structural and institutional reforms.[7] Macroeconomic stabilization measures are considered to provide the background in which other measures can effectively be taken. Indeed, fiscal policies (aimed at balancing total domestic demand with domestic production and controlling the budget deficit) and monetary policy (aimed at controlling the growth of the money supply) provide the stability that is a precondition for success of reforms. Price reforms entail the dismantling of centrally planned and controlled prices in favor of market prices that reflect the true supply and demand. The following are crucial elements of price reform: domestic price liberalization (to introduce the market as a determinant of prices), trade liberalization (to decontrol exports and liberalize imports by opening up international trade at world prices and to move trade with the USSR on a market basis) and currency convertibility. With respect to structural and institutional reforms, the following are considered essential: the imposition of hard budget constraints for firms (thus resulting in the demise of unviable firms), the stimulation of the private sector (thus diminishing the role of state ownership), the reform of the legal system (including property rights protection and tax legislation), reform of the banking and financial system (including the expansion of a capital infrastructure) and the development of a social safety net.

While these concomitants of the transition process are largely undisputed, what is under discussion is the pace of change, the extent of change, the sequencing of change and the political environment in which it takes place. With respect to the pace of change, some scholars have advocated the Big Bang approach (associated with Jeffrey Sachs[8] and Janos Kornai), while others (such as Soos and Wanniski[9]) have advocated slower, gradual change.[10] The argument in favor of rapid, all-encompassing change in the transition process seems analogous to the view expressed by Rodenstein-Rodan in the development process of less developed countries, namely that it is desirable to develop all economic sectors simultaneously.[11] So too the "Big Bangers"[12] argue for sweeping, sudden reforms simultaneously on all planes, which would cause extreme pain but for a shorter period of time than if the reforms were dragged out. Those advocating a gradual approach argue that rapid steps toward a market system would cause unbearable pain, as unemployment, decreased output, inflation and a deteriorating trade balance wreck the already precarious social and economic fabric. They argue for a sequencing of reforms, preferably the establishment of market institutions followed by the privatization of large holdings, a process that is bound to result in significant unemployment. Hence, the distinction between Big Bangers and gradualists may sometimes be reduced to a difference in sequencing.

With respect to the extent of change, again theory and practice can be divided into the radical and the conservative approaches. Among the more radical reformers, there are efforts to obliterate all elements of fifty years of communist economic measures and embark on a capitalist path, with all its trials and pitfalls. Indeed, some leaders and academics in the former Soviet Bloc have been said to be more Friedmanite than Milton Friedman. However, this approach, although popular and certainly supported by some western counterparts, is not all-pervasive in the formerly communist states. In fact, there remain advocates of a third way, a path that mixes the best of socialism with the best of capitalism—a path for the future that attempts to move to the market while avoiding the pitfalls associated with capitalism. These views, adhered to by people such as Alexander Dubchek, Mihailo Markovic, Istvan Czurska and Vladimir Meciar, were once popular in the West during the cold war (when they were welcomed as dissident views chipping away at the communist hegemony), but they are now perceived as reactionary. However, with changing world conditions, the demise of communism and the embrace of capitalism, views purporting "socialism with a human face" are perceived as irrelevant and even dangerous.[13] Western critics have said that the application of the third way will result in large economic costs and must therefore be discouraged.[14]

With respect to the political environment in which reforms occur, the crucial difference of opinion lies in the desirability and necessity of a politically free environment. The debate is between those who advocate a liberal political environment in which individual wants are communicated to the rulers and checks and balances are maintained through a multiparty system, and those who deem such a system neither necessary nor desirable. In their opinion, a proliferation of views such as a multiparty system allows, would draw attention and energy away from the goal of transition. The principal examples of these two views in action are the former Soviet Union, especially Russia, and China. In the case of the former, the glasnost reforms accompanied perestroika reforms so that the society was shaken by simultaneous political and economic changes. In China, despite the introduction of capitalist reforms in the economy, the political atmosphere was not altered—as evidenced by the policy statements reiterated as recently as October 1992 at the Communist Party Congress meeting, which clearly indicated the desirability of adhering to previous political economy policies.[15]

It is clear in 1992 that from the former socialist world a two-tier system is emerging: one is composed of countries aggressively pursuing capitalism and withstanding the pressures from the population that in theory supports market reforms but in practice is reeling from economic hardships rarely encountered during communism. These countries include Poland, the Baltic states, East Germany, Hungary and the Czech lands, in which, despite the initially discouraging results, the present situation gives reason for optimism.[16] The other tier is composed of countries whose leaders are dragging their feet, whether motivated by the desire to retain their position and popularity, or because of genuine beliefs that gradualism or "no-reformism" is preferable. In these countries, including Romania, Bulgaria and Ukraine, the reforms have been halfhearted, and the critical areas in need of change, such as privatization, currency convertibility, et cetera have been postponed indefinitely. Indeed, it is proving to be, according to Gati, easier to erect a market economy from scratch than to build one from shambles.[17] Thus, while the world laments the emergence of a two-tier system in Western Europe following the controversy associated with the Maastricht treaty, a two-tier system in Eastern Europe that will delineate future economic development in the region is clearly emerging: according to *The Economist,* "The iron curtain is gone, but a new partition of Europe is in the making."[18]

Transition in the Balkans, Excluding Former Yugoslavia

If the two-tier system is acceptable grounds upon which to divide former socialist states and their transition processes, then in the Balkans only Slovenia might be considered to be on the fast track, while Romania, Bulgaria, Albania and the remaining former Yugoslav states are undergoing halting reforms. This is a reversal of past trends, since Yugoslavia (accompanied by Hungary and Poland) previously led the way in Eastern Europe, beginning reforms in late 1989, while Romania and Bulgaria did not follow until late 1990 and early 1991. The elements of the transition process set out above will be assessed in the Balkan countries below.

In Romania, the transition consists of a privatization program, a price liberalization program and a stabilization program to offset unemployment and inflation. The privatization program is extremely ambitious, rivaled only by that in (former) Czechoslovakia. It is designed to privatize 6,200 state enterprises over a ten-year period. In June 1992, the government distributed some 4.3 million "ownership certificates" to citizens for a minimal fee. These vouchers were then considered ownership shares in five state-created investment funds and will eventually become stocks in firms slated for privatization. In April 1993, it was announced that the national airline will be the first major state enterprise to be privatized.[19] However, thus far the entire privatization program is being implemented slowly, and amid great confusion, with only one firm privatized in the first nine months of the program. Collectivized land has been restored to some six million people, although there was no concomitant law on the restitution of industrial assets.[20] Great strides were made in the development of a private sector in trade, tourism and services, as the number of private businesses grew to 362,000 by September 1992.[21] The private sector contributed: 25 percent of GDP in 1992; 33 percent of employment in 1992; 45 percent of services as of June 1993; 47 percent of retail trade as of June 1993.[22] With respect to the price liberalization program, it was first introduced in November 1990 and was followed by a deepening and widening over the course of the next three years. In May 1993, the government eliminated state subsidies for most staple goods and basic services. The ensuing price explosion was the most dramatic since price liberalization first started in 1990 (resulting in further decreases in the standard of living, widespread protests and strikes and social unrest). The government cut state subsidies in January and September 1992, pushing more firms toward bankruptcy and contributing further to unemployment. The stabilization program also included efforts to decrease the inflation rate (which it did from January to August 1992), decrease the trade deficit (which it did in 1992 by some 50 percent over the previous year). However, the measures did not succeed in increasing industrial and agricultural production or in curtailing unemployment.

Despite these planned reform programs, the international community retains doubts about Romania's commitment to reform. These doubts are amplified by the reelection (in September 1992) of Ion Iliescu to the post of president; his platform included promises to slow down Romania's economic reforms, which he directly blamed for rising inflation and unemployment. Indeed, the success at the polls of the ruling party, the Democratic National Salvation Front (DNSF), was based on its desire to "humanize" economic reforms by resisting shock therapy and opting instead for a slow and "balanced" approach and "reform that does not hurt."[23] By fall of 1993, the party changed its name in an attempt to indicate a change in course and to shed its anti-reform image, but it has yet to convince the international community.[24] Indeed, disappointment with the pace of reforms (especially in the monetary sector) led to the decision by the IMF to withhold a new loan in the fall of 1993.[25]

Bulgaria seems to have made consistent strides toward a market economy. With respect to establishing an infrastructure to support the transition, legislation authorizing the privatization of state enterprises was passed in April 1992. By then, restitution laws had already been passed, denationalizing small shops, businesses, housing and other property. In addition, collective farms were abolished. The privatization law in Bulgaria was revamped in August 1993. Previously, it was very different from that in other countries insofar as it offered the public neither vouchers nor a free distribution of shares. Engelbrekt describes the process: "State companies are to be sold mainly through auctions or public sales of shares on the stock market."[26] In mid-1993, the cabinet approved a new mass privatization plan for the sale of 500 companies, valued at $6.5 billion, in which each citizen is to receive a nontransferable option on shares worth $1,000 with which to purchase shares in a company.[27] The private sector's share of GDP was 20 percent in 1992, and its share of retail trade had reached 52.2 percent as of June 1993.[28] Other measures included the approval of a liberal joint venture law in 1992, as well as the establishment of a securities exchange in Sofia. Reconstruction and privatization of the banking system, expected to take three years, began in September 1992 with the merging of 22 commercial banks. The stabilization program that Bulgaria adopted has been under the guidance of the IMF and began in 1991. Indeed, its economic performance was recently praised by the IMF insofar as the transition was proceeding according to its prescription. Bulgaria had managed to avoid high inflation, the *lev* was fairly stable and exports exceeded the IMF's expectations.

Despite this success, the West was skeptical of Bulgaria's commitment to reform, an assessment supported by various members of the Bulgarian government.[29] This was expressed by the suspension by the World Bank of part of its loan in February 1992, on the grounds that the pace of privatization and banking reform was too slow. By spring 1993, foreign observers began identifying a "recommunization" of Bulgarian society, raising serious doubts about the future pace and extent of reforms. This was evident in the intent of the cabinet to slowly revise part of the reform legislation that was passed in 1992.[30]

Albania began its transition efforts more in an effort to arrest the sharp decline of the economy than to join a market system. The popular pressure in the state, exerted by student demonstrations in December 1990, was actually in favor of political liberalization rather than economic measures. Nevertheless, soon following the proliferation of political parties, in spring 1990, the government announced some piecemeal changes: the creation of a new economic mechanism aimed at decentralization by replacing central control and encouraging production with incentives and decentralized decision making. In addition, controls on private ownership and trade were lifted, and foreign investment was encouraged by legislation assuring all potential foreign investors.[31] These limited changes did not bring about the desired results, so they were followed by a new program of marketization, a further decentralization of the decision making process and a call for increased allocation of resources to the consumer and service sectors, as well as the combination of state and private property.[32]

During most of the 1980s, Greece was ruled by a socialist government whose policies included the creation of a large public sector and, according to some views, excessive spending relative to revenue.[33] In 1990, a conservative government came to power; it attempted to redress what it perceived to be inefficiencies of ten years of socialist rule. Its measures included public sector reform, the pullout of state holdings in firms, the selling and liquidating of public assets, the lifting of regulation of industries, the freeing of the banking system and of the labor market and the elimination of price controls. In this way, the policies associated with Prime Minister Constantine Mitsotakis in fact brought about a mild form of transformation similar to its northern neighbors. Indeed, its goal of privatization of the public sector,[34] liberalization of prices and lifting of regulation are not dissimilar to those of former Yugoslav republics.

Transition in Former Yugoslavia

Following the Soviet-Yugoslav split in 1948, Yugoslavia led East European countries with respect to the number and scope of reforms. Indeed, it had applied the concept of market socialism throughout its economy and was a pioneer in industrial democracy through the system of self-management. Its dramatic reforms of 1965 and 1974 introduced a system that was the envy of the Eastern Bloc and amazing feat as seen by the West. Its trade was highly liberalized and took place mostly with Western Europe. Western loans were forthcoming, and the standard of living rose dramatically during the 1960s and 1970s.

Economic reforms aimed at the transition to a market economy began at the federal level in the late 1980s and continued into the 1990s until the breakup of the country, setting the stage for further individual reforms. These federal reforms were conceived primarily as stabilization measures, and secondly as transition changes. The first part, focusing on stabilization, began in January 1990, while the second part, focusing on transformation, followed in July of that same year. The first program was primarily aimed at decreasing the inflation rate (which in 1989 had reached 2,700 percent; by the end of 1989 10,000 percent was the annual inflation rate). This was done with an income policy including a wage freeze and a temporary freeze in prices, as well as complete convertibility of the dinar (which was pegged to the German mark at the rate of 7 to 1). Another goal of the first program was to decrease the unemployment rate, which hovered around 14 percent (although it is estimated that surplus personnel amounted to some 20 percent of the labor force). However, efforts to deal with unemployment conflicted with efforts to bankrupt inefficient enterprises, adding to the confusion of the reforms.[35] The second part of the program was based on the understanding that stabilization policies were to be most successful in an environment of liberalization. Hence a policy of privatization and institutional changes such as a consistent tax and legal system were instituted.[36]

Prime Minister Ante Markovic came under attack from all Yugoslav nationalities for his economic policies, as he was accused of exercising favoritism among republics. Such perceptions led to unilateral action by republics aimed at undermining the federal plan, such as the introduction of internal tariffs by Serbia and Slovenia. He also ran up against the legacy of Tito in his efforts to privatize: Yugoslavia's unique form of social ownership, fostered by the system of self-management, made the process of transformation harder in Yugoslavia than in any of the other Eastern European states. Social ownership implies the necessity of transferring ownership to the state before it can be privatized. Despite these problems, the assessment of the reforms in mid-1990 was quite positive; as Slovene Pregl stated, "Yugoslavia has now passed the turning point. It is deep in an irreversible but smooth transformation or 'soft-landing' into market economy and multiparty democracy."[37]

The transitions in Croatia, Serbia, Montenegro and Bosnia-Herzegovina have been put off by the civil war of 1991-92, while those of Slovenia and Macedonia have been indirectly negatively affected. Therefore, with other national priorities, the governments had neither the capacity nor incentive to address the issues of transition. Despite the professed commitment of the Tudjman government in Croatia to proceed with radical change in the shortest possible time, there is indication that there is not consensus on this matter. According to a Croat, "When the war stops, the fighting will begin"; this comment refers to the debate about the pace and extent of reform.[38] Indeed, while President Tudjman claims to favor fast reform, including the introduction of a free market and encouragement of foreign investment, there are those in his government who fear the disruptive effects of such reform, especially foreign ownership of economic assets.[39] Croatia first introduced the Croatian dinar, with a flexible exchange rate, in December 1991. Its value is adjusted daily against a basket of European currencies in which the deutsche mark has the greatest weight. At some point in the future, after some degree of stabilization of the economy has taken place, the Croatian dinar will be replaced with a new currency, the crown (kuna) that will be pegged to the Ecu. The Croatian National Bank, the central bank, pursued a policy of tight monetary control, thus avoiding the inflationary collapse in adjoining Serbia, and it is restructuring the inherited banking system. Moreover, in March 1992, the Zagreb Stock Exchange opened in an effort to raise capital for the economy.

According to Pleskovic and Sachs, Slovenia has surpassed all former Yugoslav and Soviet republics in the transition to a market economy.[40] In the spring of 1991, the government introduced an economic program whose goal was to proceed along the path of transition while setting up independent economic institutions. This entailed a macroeconomic program that established fiscal measures (including a modernized tax system and the elimination of the link between fiscal policy and money creation), the introduction of a convertible currency and structural reform consisting of privatization and financial restructuring of commercial banks and enterprises. The problems associated with these programs, especially with respect to privatization, have, according to Zapp, been insurmountable due to political disputes.[41] With respect to the monetary changes, aimed at both establishing sovereignty over its economy and insulating the economy from the inflationary pressures from the rest of Yugoslavia,[42] the following was done: a central bank was created to introduce monetary and exchange rate policies; legislation was adopted to establish a market for foreign exchange and thus eliminate the black market; the tolar was established as legal tender, convertible on a one-to-one basis from the dinar, against which it would float freely, as with international currencies, in October 1991.

Macedonia has been unable to make great strides in the transition process to a market economy. Preoccupied with the disruption of markets and financial flows due to sanctions and the breakup of Yugoslavia, coupled with the Greek blockade on its southern border, Macedonia has adopted reforms including the introduction of the republic's new currency (the denar), an anti-inflationary program, new legislation on fiscal policy and trade and the preparation of new legislation on the restructuring of state-owned enterprises and privatization.[43] The first step in the privatization process was the privatization of state-owned apartments, since this was expected to yield considerable foreign currency.[44] Progress on price liberalization is slow, as the government had first to deal with the isolation of its currency from the Yugoslav dinar. In this effort, it separated its currency and devaluated it in October 1992 by 66 percent. At the same time, some prices were liberalized (such as sugar and cooking oil), while others were adjusted upward (electricity, municipal services).

The federal program for the transition of former Yugoslavia was halted in Serbia/Montenegro because of the war and altered dramatically following the introduction of sanctions. The economic policy of the new Yugoslav government, as adopted in 1992, consists of three facets: a program of stabilization, focusing on anti-inflation measures, a program to develop a market economy and a program for the transition of the economy.[45] The stabilization program includes monetary and fiscal policies, currency reform, wage and price policy and changes in efforts to secure outside financial support. The development of a market system includes the development of market institutions, such as banks and a stock exchange. The program for the transition includes changes in businesses, the banking system and social programs. Yugoslav policy with respect to inflation is aimed at checking hyperinflation and market mechanisms are attempted everywhere (even in the military).[46] All of these measures are adopted so as to be sustainable under a system of sanctions and war economics, as described below. They nevertheless were met with severe criticism from President Milosevic's opposition, especially from the Serbian Radical Party, whose commitment to economic reforms is not extensive.

WAR ECONOMICS

War and the Economy

Numerous scholars have written about the relationship between war and the economy from various perspectives, including the pursuit of war for economic reasons,[47] economic effects of war in the specific case of capitalism[48] and the relationship between war and economic functioning and development. This last perspective is relevant to this study.

The relationship between war and the economic functioning of a society has been a source of much debate. Barbera has offered four basic possibilities: "War is either productive, destructive, unpredictable or irrelevant" to economic growth and development.[49] The first two of these are more common; in the least, war affects economies insofar as it necessitates the choice between "guns and butter." Despite differences in war potential among belligerents, resources available for war (such as raw materials, food and manpower) and the location of war, all sides experience the ramifications of transforming a civil economy to a war economy. Neal has identified some of the principal ramifications of war for the economy, and these are adopted in the discussion of the Yugoslav war.[50]

1. *War production:* One of the primary changes in the economy is the shift from civilian to military production, in order to make more war products available, especially iron, steel, and aluminum, as well as food and clothing.

Clearly, the diversion of these goods from their prewar destinations in the production process or as final goods has an impact on the economy. The extent of the diversion of the region's productive resources depends upon the existing state of armaments, the state of the enemy's armaments, the length of time before they will be needed and the self-sufficiency of the economy (specifically, the availability of raw materials and industrial goods). The techniques used to bring about this transformation depend upon the preexisting nature of the economy. In a market economy, an alteration of prices may be used to bring about a change in production. However, if the time is short, or if the economy is of a command nature, then prices may be set by the government. The government may also simply buy up a needed input, prohibit its use for other things, establish a system of licensing, or simply take over the source of supply. The production of war materials may be expanded, both by adaptation of existing facilities and by the construction of new ones. As shortages and bottlenecks develop, governments are more inclined to use direct controls to expand military output.
2. *Displacement of competitive processes, if they existed*: If there was preexisting competition in the economy, it is of necessity disrupted. Government requirements cause disruption of the normal process of price determination as the government enters the markets and sets prices to alter production and distribution.
3. *War labor problems:* During wartime, numerous demands are placed upon the labor force. First, mobilization requires the withdrawal of people from routine economic activities and their placement into war activities. Therefore, the supply of labor decreases. Second, increased military production may increase the labor requirements in those industries. Thus, labor market disequilibrium occurs as the demands for labor increase while the supply of labor decreases. This has two results: first, boosting of wartime morale to draw volunteers (such as women and retirees) into the market, and second, a conscious wage policy that necessarily differs from a peacetime policy.
4. *Financing the war effort*: The war effort may get financed from public or private sources. The former entails the raising of taxes and the floating of loans, with the particular conditions dictating to what extent these activities will be engaged in. The latter entails the use of business capital, as provided by owners and investors, commercial banks and internal financing.
5. *Management of monetary and banking systems*: This entails the direct control of prices, which are under inflationary pressure due to increased government demand for goods and subsequent shortages. The alleviation may also entail rationing and the restriction of personal consumption. During wartime, the banking system must exhibit flexibility and bend to government

pressures. It is extremely difficult for the government to set the supply of money and credit during war.

6. *Economic warfare*: This aspect of war economics entails measures designed to lessen the economic strength of the enemy by preventing it from getting supplies of important commodities: such measures include a boycott of trade, financial restrictions and a naval blockade.

War and Economics in Yugoslavia

The civil war in former Yugoslavia has brought on billions of dollars of damage (estimated at $20 billion in Croatia in 1992),[51] devastated lives, created some two million refugees and resulted in a major economic setback for the entire Balkans region. It broke out in earnest in July 1991 in Croatia, and by spring 1992, the location of the fighting had shifted to Bosnia-Herzegovina, where it proved to be more devastating than in Croatia, not in the least because there were now three groups of enemies with ambiguous alliances. The waging of these wars entailed changes in the economies of at least Croatia, Serbia/Montenegro and Bosnia-Herzegovina. While the war did not take place in Serbia proper, the region is nevertheless included in this study because it exhibits characteristics of a war economy: war is being waged at its borders, aid is being siphoned off to Serbs in the warring zone, and refugees are a burden on the economy. The internal markets of former Yugoslavia have been disrupted and sanctions have disrupted the external markets. The economy of Bosnia-Herzegovina does not exist as such: rather, there are a multitude of virtually self-sufficient local economies coexisting with minimal trade and minimal aid and no central infrastructure tying them together.

War Production

The Yugoslav wars caused production alterations in neighboring Macedonia and Slovenia, as these two states adjusted to the disruption of Yugoslav markets. The Slovenian economy, which was based on the purchase of raw materials (at low prices) from the Yugoslav markets and the sale of its manufactured goods (at relatively high prices) to these same markets, suffered a 30 percent decrease in production as a result of the disruption of trade. Macedonian production was altered as a result of the disruption of markets and trading, but also for another reason: when sanctions were first imposed on Serbia, the border with Macedonia proved to be porous and Macedonia's economy adapted to fulfill the new demands from Serbia. Thus the production of manufactured goods as well as some food was increased to satisfy the new export demand. When sanctions were then strengthened and U.S. troops were sent to guard the border to ensure compilation with the sanctions, the government of Kiro Gligorov expressed concern and urged the United States to send aid to the Macedonian economy in lieu of soldiers.

However, it is the Serbian and Croatian economies that have experienced the greatest production shifts in order to accommodate the war. The Serbian government made a plan, at the time of the imposition of sanctions, to exert state control over a large number of economic sectors. While the government efforts in Serbia are described in detail in chapter 5, suffice it to say here that they involved austerity measures and government control over key sectors in the economy (food, medicine, energy, etc.), and the accelerated printing of money. This was necessitated in part by the demands of the war on the economy, and in part by the demands of the sanctions. In Croatia, the loss of production has been somewhat compensated for by extensive aid as well as the imports. Nevertheless, the Croatian government became increasingly involved in ownership and control of crucial branches of the economy.

Displacement of Competitive Processes

There has been a mixed result with respect to the displacement of competition by the war. While on the one hand there is no doubt that the governments of both Serbia and Croatia have extended state control of various aspects of their respective economies, there has also been a proliferation of competition on a different level. Increased state control is evident in the renationalization of firms and banks. Croatia enacted a privatization law in 1991 that enabled managers and workers' councils in socially owned enterprises to choose between several privatization schemes. However, this was never implemented as a result of the war, and in fact, privatization was replaced with government control over principal sectors. In Serbia, there has been widespread debate on the benefits of privatization. While the privatization process was started with legislation in 1990, some have claimed that its only result has been the pillage of social property.[52] By mid-1993, some 20 percent of social ownership was privatized, leading Bozidar Cerovic to call it "privatization by the spoonful."[53] Thus far, it seems that the war is leading governments to introduce command economies of the kind that Yugoslavia only experienced in the immediate aftermath of World War II.

At the same time, there has been a growth in minor private activity, mostly in the services but also in small-scale production. This has been tolerated and even encouraged in both states because it takes up the slack from the government at a time of great shortages. Indeed, the private supply of items such as medicine and gasoline has greatly reduced the pressure on the government to provide them.

War Labor Problems

The Yugoslav wars have had a negative effect on the labor force, mostly due to the decrease in the supply of labor. This has occurred for several reasons. First, many workers have been drawn into the conflict, leaving their enterprises short of skilled labor. Shortages like the one at Croatia's principal shipyard, which is 600 people short of its 5,000-person workforce, impeding the functioning of the unit,[54] are common occurrences throughout the state. In addition, the creation of a new army and police force has removed numerous workers from the labor market: a low estimate of Croatia's labor shortage stands at 200,000 workers, not counting the irregular volunteers. Second, the supply of labor has also decreased because of the emigration of workers out of the warring states, in order to escape both a possible draft and dire economic conditions. Very significant has been the exodus of the skilled population (brain drain), which has affected Serbia especially harshly. (It is estimated that some 100,000 to 150,000 professionals left Serbia in 1992.)[55] Third, the war has created refugees, as people have lost their homes or have been "ethnically cleansed" from their homes. These population movements have caused shortages of labor in some locations, and an overabundance of labor in others. Fourth, given the deteriorating demand and economic production, workers have been laid off. Together these forces affect unemployment, which in Croatia amounted to 20 percent of the labor force in early 1992. In Croatia, a new program for invigorating the economy was introduced in April 1993, employing 30,000 in a public works program to repair the war-damaged infrastructure.[56]

Financing the War Effort

In Croatia, a tax increase was introduced to fund the war; the increase was so burdensome that the government was obliged to reduce it for lower-income households in 1992. In addition, the demands of war on the state budget were so high that in the last trimester of 1991, 90 percent of the budget was allocated to defense purposes.[57] However, it is hard to estimate the true numbers since the military expenditures are not listed as a separate item in the budget.[58]

In Serbia, the budget for 1993 was altered to include an allocation of 75 percent to the military.[59] In addition, the government froze individual and enterprise foreign currency holdings, which it "borrowed" to finance the war effort. Estimates of this amount are hard to come by, although one source puts it at an unbelievable $12 billion in private hard currency accounts.[60] The Yugoslav government authorized the periodic printing of money to finance the war effort until it was stopped by Prime Minister Milan Panic in the summer of 1992. Moreover, Serbia announced a moratorium on repayments of its foreign currency debt, in part because of the need to retain the foreign currency reserves for the war.[61]

Management of Monetary and Banking Systems

In Serbia, hyperinflation has exceeded all textbook descriptions. As a result, barter has increasingly become the preferred method of exchange. The dinar has been replaced by the deutsche mark as the currency in which all prices are quoted. These phenomena are also evident in Bosnia-Herzegovina, where five currencies are in use, and to a lesser extent in Croatia.

Economic Warfare

Before actual fighting broke out in the former Yugoslav republics, economic war was raging. One example of this is when in mid-1990 the Serbian Republic imposed duties on goods from the republics expressing a desire to secede, Slovenia and Croatia. This move was immediately followed by retaliation on the part of the Slovenes and together they had the effect of abolishing the internal common market of Yugoslavia. At the same time, propaganda encouraging boycotts of goods produced in other republics was conducted in all republics.

The economic effects of the Yugoslav civil war and the concomitant government responses can be compared in nature but not extent to similar actions during the period of war communism in Soviet history. During this time (1918-21), the Soviet leaders adopted policies that were, in the opinion of Dobb and Carr, forced upon them by the civil war.[62] The goal of these measures was to substitute administrative allocation of resources for market mechanisms in order to provide the necessities of war. This entailed the forcible requisitioning of food surpluses, the nationalization of the economy, the abolition of private trade (which created an enormous black market), the use of semimilitary controls as a means of labor allocation and a class system of distribution (rationing). In addition, the use of money was virtually eliminated because of hyperinflation and was replaced by a system of barter and physical allocation. Some of these measures were attempted in the war-ravaged economies of former Yugoslavia for the same reasons as they were in the Soviet Union.

THE CONFLUENCE OF THE ECONOMICS OF TRANSITION AND WAR

At the beginning of the Yugoslav civil war, political, security and military issues began to dominate in former Yugoslavia and had a trickle-down effect even in regions not directly affected by war. Thus reform pertaining to the transition from socialism to capitalism became relegated to second place. Indeed, various aspects of the transition process were in direct contradiction with the demands of a war economy, and thus certain goals became mutually exclusive. In the case of the new Yugoslavia, it is not the war that is principally responsible for the destruction of the economy but rather the application of international sanctions and domestic policies (described in chapter 5). Research provides ample evidence that the principal reasons for the deterioration of the economy and the lack of economic reforms are the imposition of sanctions and the loss of Yugoslav markets rather than the war per se.[63]

As previously noted, three categories of change are associated with the transition process. How transition is affected by war is discussed next.

1. *Macroeconomic stabilization measures*: Fiscal policies aimed at controlling the budget deficit are in direct conflict with the need to finance the war effort. This is clearly exemplified by both Serbia and Croatia, where the proportion of the state budgets devoted to military affairs has risen rather than fallen over the past two years. The control of the growth of the money supply is also in conflict with the demands of a war economy. Indeed, the printing of money by the Serbian treasury does not deflate prices. In order to offset the effects of the war (and sanctions), Popovic predicts the establishment of a command economy "of the Cuban type" as all other measures of stabilization are incompatible with a war economy.[64]

2. *Domestic price liberalization, trade liberalization and currency convertibility*: These aspects of the transition reforms have all been discontinued in Croatia and Serbia. Under conditions of rampant inflation, price liberalization has been discontinued, and instead price controls have been introduced for a variety of goods. Rationing—especially of flour, sugar, oil and detergent—has also sporadically occurred. With respect to opening up to trade at world prices, only Croatia has been able to engage in any form of trade.

Serbia has been paying prices higher than world prices for its imports because they have been entering the country under conditions of sanction busting (see chapter 5).
3. *Structural and institutional reforms*: The issue of privatization in the war-torn Yugoslav republics is complicated by the following two factors. First, strategically important companies have been taken over by the state, in an effort to control production. Also, in both Serbian and Croatian agricultural production, there has been forced procurement of foodstuffs and raw materials used as inputs in further production. Second, companies based in one republic with branches in another were taken over (or bought) by the government of the host republic. Bicanic said that these companies had been "regionalized"; this occurred in Slovenia, Croatia and Serbia.

Given the confluence of pressures from the transition process and the war, and the relative immediacy of the latter, one cannot but worry about the future of some of the liberalization trends in both Croatia and Serbia. In looking forward to a time when the war ceases to make demands on the economy, the Yugoslav leaders cannot but be reminded of the Soviet experience in the aftermath of war communism: In 1921-28, in order to rectify the problems resulting from the highly unpopular policies of the war, the New Economic Program was introduced, in order to stimulate long-term growth. This program had as its cornerstone the reestablishment of the market economy, the stimulation of the agricultural sector, the reestablishment of private trade and the decentralization of decision making at the micro level. Some nationalization was reversed, foreign trade was encouraged and the reintroduction of the use of money was a priority. These measures are not unlike those found in the reform packages of numerous formerly socialist countries on the transition path to a market economy. Might the analogy stop there, or might there be a massive industrialization debate (such as the Soviet industrialization debate of 1924-28) and a Stalin lurking in the future of the Balkans?

CONCLUSIONS

Two factors will have a large impact on the transition process in the Balkans, as well as on the reconstruction abilities in the aftermath of war and sanctions. These are internal political change and foreign investment. The role of political change as a background to economic transformation has been clear in numerous former communist states. In the late 1980s, it was common to focus on how glasnost preceded perestroika in the Soviet Union, while in China, perestroika preceded glasnost, thus underscoring how the political and economic reforms do not necessarily coincide or occur in a specific order. In the 1990s, political support for economic reforms became crucial in numerous states undergoing transition. Indeed, in Romania, Russia and Serbia, recent elections were focused on reforms and the economic effects of transition. The divergence in views pertaining to the pace and extent of economic reforms was evident in Romania between the Democratic National Salvation Front and its chief contender, the Democratic Convention Coalition;[65] in Serbia between President Slobodan Milosevic and Milan Panic, his rival in the December 1992 elections; and in Russia in the battle between Boris Yeltsin and members of the parliament in the April 1993 elections. Clearly, without political change to support them, reforms will have difficulty even being implemented. On this score, Gati said, "In the absence of a more tolerant political culture, chances for economic progress are poor; without economic progress, chances for the new democracies to take hold are equally poor."[66] Numerous economists have argued that the hostility to change and inherent instability in the political atmosphere explain why the former Eastern Bloc countries will not "bounce back" as easily as some South American states did, despite their similar economic problems.[67]

The slowness and even reversal of reforms that characterize the transition in the Balkans is bound to have a further negative effect on the region, especially on private and government foreign investment in the region. There is no doubt that an influx of capital could play a pivotal role in the Balkans' transition at this time, by enabling reforms to succeed as well as by repairing the war-ravaged infrastructure. However, private investors have been scared away from the region, suspecting more war, nationalization and controls. International organizations such as the IMF withdrew support pending the implementation of reforms to their liking. There is no doubt that foreign investment in the Balkan regions can only succeed if privatization and other reforms have taken root. Without progress in reforms, countries simply cannot absorb large quantities of external funds. Indeed, *Euromoney* recently concluded an assessment of 169 countries on the basis of their investability, which they called a "lendscape." This review ranked the Balkan states in the middle range, with the exception of Albania (see table 4.1). The results are supported by a poll of international investors conducted in mid-1993 in which chairs of companies were asked where they have invested in the recent past and where they plan to invest in the near future.[68] The results indicate that Slovenia can expect significant increases in investment, Romania and Bulgaria can expect little and Serbia and Croatia can expect none.

Table 4.1
Balkan "Lendscape"

State	Ranking (1992)	Ranking (a) (1991)	Total (out of 100)	Economic Performance (out of 10)	Political Risk (out of 20)	Debt Indicator (b) (out of 10)
Romania	72	89	35.8	4.0	7.2	9.1
Slovenia	74	—	34.2	4.1	7.9	9.2
Bulgaria	911	14	29.9	2.8	5.7	8.9
Croatia	101	—	26.5	2.9	4.4	9.2
Macedonia	108	—	25.1	1.4	4.0	9.2
Yugoslavia	125	—	22.7	2.0	1.5	9.2
Bosnia-Herzegovina	131	—	21.3	0.4	1.7	9.2
Albania	142	125	18.5	2.4	2.1	0

(a) In 1991, the ranking included unified Yugoslavia, not the individual republics.

(b) The debt indicator score is calculated from the following: debt service to export ratio; current account balance to GNP ratio; external debt to GNP ratio.

Source: *Euromoney*, September 1992.

5

Sanctions: Trigger of Economic Collapse or Stimuli for Economic Initiative?

"A nation that is boycotted is a nation that is in sight of surrender. Apply this economic, peaceful, silent, deadly remedy and there will be no need for force. It is a terrible remedy. It does not cost a life outside the nation boycotted, but it brings a pressure upon the nation which, in my judgement, no modern nation could resist."

Woodrow Wilson[1]

"In some cases, domestic political goals were the motivating force behind the imposition of sanctions. Such measures often succeed in galvanizing public support for the sender government, either by inflaming patriotic fever (as illustrated by U.S. sanctions against Japan just prior to World War II) or by quenching the public thirst for action (as illustrated by U.S. sanctions against Libyan leader Moammar Qaddafi)."

Gary Clyde Hufbauer et al.[2]

"Yugoslavia boasts the most porous borders in Europe thanks to a conspiracy of geography and corrupt neighbors."

Misha Glenny[3]

Sanctions constitute yet one more area in which policies of Woodrow Wilson have both opened and closed this century. Indeed, not only is his involvement in the redrawing of Balkan boundaries relevant today, but so is his support of the use of sanctions to alter the behavior of targeted states. As with other fashions, the popularity of sanctions as a policy tool has waxed and waned throughout history. They were popular after World War I as multilateral measures taken through the League of Nations and then again during the cold war as unilateral measures taken largely by the superpowers to control states within their spheres of influence. Today, after the cold war, once again sanctions are being waged as a tool to obtain a desired effect. Despite the numerous cases of sanctions in which one power is the sanctioner (such as the ongoing U.S. efforts against Cuba), currently the principal cases involve a group of states operating through the United Nations against a target state—for example, Iraq and the new Yugoslavia. Such unprecedented unity on the part of the world body was not possible in a bipolar world, so it is likely that the use of sanctions will grow in the near future. However, unity of sanctioners in no way implies that either the success of sanctions or world peace will grow.

SANCTIONS IN THE BALKANS

Economic sanctions are defined as the cessation of trade and financial relations as a result of governmental action. Economic sanctions include the limitation of exports, restriction of imports and a prevention of financial relations. They can include the freezing of assets abroad and the isolation of the country with respect to travel, communications such as mail and telephone, and so on. The purpose of sanctions is to affect a target country in the following ways: alter policies (for example, human rights, terrorism or apartheid, as in the case of South Africa), destabilize the government (such as U.S. efforts against Fidel Castro or Saddam Hussein), alter the outcome of a military adventure (such as the British efforts against Argentina over the Falkland Islands) or reduce the military potential of a belligerent state (such as sanctions imposed during the world wars). The goals that underlie sanctions are more easily achieved under certain conditions. These conditions have been enumerated and studied by Hufbauer, Schott and Elliot, and are defined by political and economic variables.[4] Political variables include companion policies (such as covert or military activity), the number of years economic sanctions were in force, the extent of international cooperation, the presence of international assistance to the target country, the political stability and economic health of the target country and the warmth of prior relations between sender and target countries. In the economic sphere, the following factors are included: the cost imposed on the target country (measured in terms of absolute, per capita and percent of GNP), commercial relations between sender and target countries measured by trade between them (as a percentage of target country's trade), the relative economic strength of the countries (sender and target), the type of sanctions used and the cost to the sender country. Each of these will be studied below in the particular case of the new Yugoslavia.

Sanctions are not new to the Balkan states. From World War I until 1992, there have been eight instances of sanctions, three of them involving the Yugoslav lands. In the period after World War I, the League of Nations ordered sanctions twice. Sanctions were imposed against Yugoslavia in 1921 with the goal of retaining the 1913 borders of Albania and blocking Yugoslav attempts to take territory that was allocated to Albania by the London Conference of Ambassadors. Also included were sanctions against Greece in 1925 forcing Greece to withdraw from occupied Bulgarian border territory. In the post-World War II period, sanctions were used in numerous unilateral efforts to alter policy in Balkan states. During 1948-55, the Soviet Union imposed sanctions on Yugoslavia to force it to rejoin the Soviet Bloc and to destabilize the Tito government. The Soviet Union had similar political goals in Albania, when it imposed sanctions in 1961-65 to destabilize the Hoxha government in retaliation for its alliance with China. Romania was also the target of Soviet sanctions as a result of its economic independence (1962-63). In 1975, the United States imposed limited sanctions on East European states to force a liberalization of Jewish emigration. China also saw fit to impose sanctions in the Balkans, punishing Albania for its anti-Chinese sentiment in 1978-83. Most recently, the United Nations imposed sanctions in 1992 on the new Yugoslavia for its involvement in the Bosnian war.

Four of the sanctions imposed on the Balkans in the post-World War I period were invoked because of border issues: three involved the Yugoslav lands and one involved Greece. All of these also involved the world community through an international body, either the League of Nations or the United Nations.

SANCTIONS IN THE NEW YUGOSLAVIA

On May 30, 1992, the United Nations, under the instigation of the United States, voted to impose sanctions on the new Yugoslavia under UN Security Council Resolution 757. These sanctions were overwhelmingly supported by members of the United Nations, including Russia, the traditional Serbian ally, which first opposed and then supported their imposition. The purpose of these sanctions was primarily to influence the outcome of the war in

Bosnia-Herzegovina, to "take a stand against aggression"—to send a message that borders cannot be changed by force, to punish perpetrators and to aid in the toppling of the Milosevic government, which is viewed as responsible for the war.

There was another, less overt, purpose to the imposition of these sanctions, one that had at its core the internal politics of the United States and, to a lesser degree, of Western European governments. The Bush administration had been under pressure from various interest groups across the United States to "do something" in Bosnia. This feeling was accentuated by the media, which brought pictures of devastation and death into living rooms, creating a popular surge of demands for action, any action, to remove those pictures from their lives. Last but not least was the pressure from presidential candidates, who accused the Bush administration of inaction. The combination of these pressures required that the administration take a strong stand, and thus pressure was exerted on the members of the Security Council to vote in favor of sanctions. Thus the public was satisfied, a moral statement was made and time was bought for the government to decide future moves, such as military intervention. This was not the first time that the imposition of sanctions served domestic political purposes. Indeed, previous instances of such behavior on part of both U.S. and Soviet leaders are documented, as is the famous retort by Lloyd George in reference to the League of Nations sanctions against Italy in 1935 for its involvement in Abyssinia, "They came too late to save Abyssinia, but they are just in the nick of time to save the Government."[5]

The text of the UN resolution imposing sanctions on the new Yugoslavia included the following measures: all trade in all commodities was to cease; all foreign assets of the new Yugoslav federation were to be frozen; all air traffic was to be suspended; all repair, service and insurance for the aircraft of Yugoslavia was to be arrested. After some six months of these sanctions, in November 1992, during which evidence emerged that the sanctions were not having their desired effect (as a result of sanction busting aided by neighboring states), the United Nations voted to tighten the embargo in the following ways: First, it empowered monitoring ships to inspect cargo passing into the Adriatic Sea and the Danube River and to interdict those vehicles that engage in sanction busting. Second, cargo vehicles were banned from crossing the country with oil, machinery, steel, tires and other critical goods. With UN approval, some of these rules could be bent, as occurred when the opposition parties in Yugoslavia were granted the right to import several million dollars' worth of media equipment for broadcasting in anticipation of the upcoming December 1992 elections. In April 1993, again under U.S. sponsorship, a further tightening of sanctions was approved by the United Nations, and these were put into effect on April 30 when the Bosnian Serbs refused to sign the Vance-Owen Peace Plan. These new measures included prohibiting the transport of any goods through Serbia and Montenegro; prohibiting the entrance of any ship into former Yugoslav rivers or sea; the restriction of truck and train passage to a few border points; and the seizure of all ships, freight trucks, trains and aircraft outside of the country. Trade was banned for all goods except medicine, food and relief supplies and services. In addition, public and private financial assets in foreign banks were frozen. Then, in October 1993, the UN Security Council adopted a new resolution linking the lifting of sanctions in Yugoslavia to the resolution of the conflict in Krajina, thus extending the original mandate to encompass a wider solution to the Yugoslav conflict.

The sanctions against the new Yugoslavia extend into the political sphere as well. Yugoslavia suffered from a lack of international recognition. While other states born of the former Yugoslavia have received swift recognition (notably, Slovenia and Croatia in the few days following January 15, 1991), the new Federal Republics of Yugoslavia have to date not been formally recognized, with the exception of a de facto recognition by China, Russia and Romania. Moreover, the new Yugoslavia, composed of Serbia and Montenegro, was not automatically granted the Yugoslav seat at the UN in the way Russia assumed the Soviet Union's seat in January 1992. Instead, the Security Council passed a resolution on September 22, 1992, that Serbia and Montenegro will not inherit the Yugoslav seat and will have to apply for membership in the future. Yugoslavia was expelled from the United Nations' principal bodies on September 22, 1992, by an overwhelming vote of 127 to 6, with 26 abstentions.[6] On April 28, 1993, the Security Council voted to drop Yugoslavia from the UN Economic and Social Council. While the Yugoslav experience with respect to the imposition of sanctions is not unique, its expulsion by the United Nations is unprecedented: indeed, the UN had previously suspended South Africa's voting rights in the General Assembly and imposed partial economic and partial diplomatic embargoes on Iraq and Libya, but it did not expel them. This diplomatic isolation of the new Yugoslavia extended to the closing of embassies of most western states, coupled with the withdrawal of their ambassadors. The tightening of the sanctions in April 1993 further reduced the staffs at diplomatic missions. Furthermore, it entailed Yugoslavia's suspension from the Conference on Security and Cooperation in Europe (CSCE) in June 1992 for a three-month period, which was extended indefinitely in October.

In the cultural sphere, international organizations have been instructed to suspend Yugoslav athletes from sporting events (Yugoslav participants to the Olympics in Barcelona in 1992 were allowed to participate as individuals but not as representatives of their state), exchange programs were terminated and cultural and scientific links were suspended.

Economic Effect of Sanctions on the New Yugoslavia

After one year of sanctions (mid-1992 to mid-1993), the loss of revenue in the new Yugoslavia is estimated to be some $25 billion, and the per capita national income has dropped by an order of ten, from around $3,000 to $300. In that one year, the price of bread has increased 800 times, while the price of milk has increased over 1,000 times. GNP dropped by $12 billion in that year, the value of foreign trade fell by $9 billion, industrial output fell by 40 percent in the first five months of 1993 over the same period in 1992 and one-half of the labor force is unemployed.[7] According to the Belgrade Economic Research Center, 97 percent of the population is living at the poverty level.[8] Moreover, it takes three and one-half monthly salaries to purchase the same bundle of goods that could have been purchased with one month's salary in 1990.[9] The degree of underdevelopment that occurred over a short period of time was described by a Belgrade reporter as a movement from a "rent-a-car" society to "rent-a-horse" society.[10] In Montenegro alone, the sanctions are responsible for a loss in revenue of $277 million: businesses have lost $130 million in exports, $90 million in tourism, and $57 million from shipping.[11]

The manifestations of economic decline in the aftermath of the imposition of sanctions were not immediately apparent to the casual observer. Indeed, despite the initial confusion in the markets and short-term shortages, adjustments were quickly made. Hinic and Bukovic point out that within a few months of the imposition of sanctions, economic activity increased; in fact, the dropoff was most dramatic in the month of June 1992, and industrial production picked up already by July, while trade did not pick up until mid-October.[12] The source of this activity was both public and private: the government revealed its stockpiles of goods, which it brought to the market, the population had reserves of foreign currency that it began using for daily survival as well as speculation in some shaky but profitable newly emerging banks and porous borders enabled easy sanction busting. All this led to a ludicrous overabundance of imported goods, including gasoline—so much so that the joke that made the rounds in the fall of 1992 was: Question: When will Belgrade blow up? Answer: When someone lights a match.

Over time, however, the effects of the sanctions became evident to the naked eye. After 15 months of sanctions, the author clearly noticed a decrease in automobile traffic, empty shelves in stores, unreplaced lightbulbs, long lines for milk, dealers in foreign currency producing a distinctive "ssssss" sound as they uttered under their breath "*devize devize*" ("foreign currency"). These clear indications of a deteriorating economy are substantiated by the statistics that indicate increasingly the toll on the economy: For example, numerous large factories have been closed (as industrial production has been cut by 46 percent, retail trade has been reduced by 57 percent and exports by 69 percent),[13] numerous people are unemployed (as hundreds of thousands of workers are receiving reduced wages or none), the inflation rate is soaring and shortages are spreading across sectors and geographical areas. Large industrial concerns, such as the textile manufacturer Kluz, have laid off thousands of workers and closed down because their output was aimed mostly for exports. Indeed, by the time sanctions were in place for ten months, cotton production was at only 15 percent of capacity, while this number was 20 percent in knitwear and 40 percent in the garment sector. This is due in part to the lack of imported raw materials, and in part to the lack of external markets to sell to.[14] Most construction projects have been suspended. Fuel and energy have been in short supply, as imports had ceased and the preexisting Yugoslav power stations were too large to operate for the diminished Serbian and Montenegrian markets. By mid-summer 1992, lines for gasoline extended for seven days and purchases was made with coupons.

In early 1992, food was still abundant because of the region's rich agricultural land and two consecutive good harvests. However, a drought in 1992 caused a drop of 22 percent in agricultural output. The government promised that three million tons of crops would be harvested after the 1993 harvest (based on the successful sowing of 850,000 hectares), amounting to twice the annual consumption.[15] However, this great a harvest was unlikely since Yugoslav agriculture is dependent on modern technology, and sanctions have prevented the use of sufficient fertilizer, pesticides or fuel for mechanical inputs that are necessary to accompany the seeds (or, for that matter, sufficient oxen and horses that could replace mechanical inputs). It turned out that the crop of 1993 was again a disappointment, leading people to complain that "nature, too, has colluded with the international community against Serbia." All this led to the introduction of food rationing in September 1993 among the 10 million people that make up the population, as well as the 600,000 refugees from Croatia and Bosnia. Coupons entitling households to 6 kg of flour, 500 gm of sugar, 250 gm of salt, 3/4 liter of cooking oil, and 500 grams of laundry detergent per month were distributed. These quantities, while they represent a decrease in food consumption compared to 1990, nevertheless represent an improvement over the summer 1993 months for many urban residents. The government has been quick to point out that food rationing also occurred after World War II and lasted until the mid-1950s. However, the media have been merciless: it was pointed out that the bundle of goods available in the postwar years was greater than that available now.[16]

While the population of Yugoslavia will not starve in the short run, other humanitarian pressures will emerge, such as the need for heating and medicine. While insufficient heating oil comes into the country, it is estimated that numerous households will simply not have access to or be able to afford heat. Indeed, Ramet suggested that the most immediate effect of sanctions would be in the sphere of oil, since Serbian oil fields produce only 25 percent of the country's needs.[17] An unverified claim by Croatia is that 400 tons of oil are pumped daily from an oilfield in Serbian-controlled Slavonia.[18] Even these levels of production cannot be sustained if foreign contracts for the importation of equipment are cancelled. Serbia's supplies of rationed gasoline to the residents of Belgrade decreased from 1,000 tons per day to 100 tons per day.[19] In expectation of a gruesome winter in 1993-94, the government issued a statement during the summer of 1993 urging the urban populations to move in with relatives during the winter in order to conserve heat. This was followed by an announcement from the Serbian state-run petroleum industry that urban housing will not be able to be heated above 5 degrees Celsius, as the country is running out of oil.

Similarly, the sanctions have resulted in a humanitarian disaster in the making; fundamental health care is being denied the civilian population as a result of the lack of medicines and medical supplies. X rays are not done, anesthesia is unavailable, antibiotic supplies are dwindling and vaccines for children are in shortage. As a result, nearly 50 percent of Belgrade's schoolchildren are anemic, 26 percent of the children in Nis are undernourished, and 17 percent of army recruits are rejected because of undernourishment.[20] Moreover, deaths from contagious diseases have increased by 5.4 percent over 1992, and the average calorie intake in Novi Sad has dropped from 3,200 to 2,100 daily.[21] While the United Nations will listen to special requests, obtaining exemption certificates is a long and cumbersome process rarely taken advantage of. As a result, state-owned pharmacies are empty while private pharmacies sell products at exorbitant prices, making it difficult to procure medicines. The situation has so deteriorated that the World Health Organization (WHO) has appealed to the UN Sanctions Committee about the precarious health situation and urged them to take steps to allow pharmaceutical companies to import raw materials and equipment for the production of medicines. While WHO officials requested some $43 million in relief aid, only 11 percent has thus far been collected.

In addition to the humanitarian effects of the sanctions, numerous aspects of the economy are affected. For example, the military is deteriorating; there is evidence that weapons and vehicles are falling apart without spare parts. Another area in which decline is evident is tourism. As discussed in chapter 3, tourism was an important source of revenue, both in domestic and foreign currency, that has totally dried up in the past year. The coastline of the Montenegro attracted foreign and domestic visitors, while skiing in the interior was a favorite pastime for urban Serbs. The unstable situation and the lack of money brought tourism to a halt. Moreover, the general apathy and depression of the population experiencing such a dramatic decrease in the standard of living is translated into a decreased productivity of the employed labor force.[22] Under conditions in which many hours of the workday must be spent scrounging for food, waiting for coupons and the withdrawing of money from bank accounts, it is not unreasonable to expect labor productivity to have decreased.

But the most serious and perplexing aspect of the sanctions is the effect that they have had on prices. It is an understatement to say that the value of money has dropped dramatically. At the onset of sanctions, 5,000 dinars had a value of $550, while three weeks after sanctions were imposed, their value dropped to $2.70.[23] In the course of three weeks in August 1992, the value of the dinar had dropped from 9 million per deutsche mark to 35 million. This perpetual depreciation of the dinar led to the adoption of the deutsche mark as the legal tender, as prices for most goods are simply quoted in marks. By May 1993, one year after the imposition of sanctions, inflation was at 205 percent monthly (84 million percent annually). In August 1993, it rose to 1,880 percent (at an annualized rate, it was 363,000,000,000,000,000 percent).[24] Inflation only subsided somewhat in September 1993 after the government imposed price controls.

In measuring the effect of sanctions on the Yugoslav economy, it has been difficult to identify how much of the economic disaster is the result of sanctions, how much is the result of mismanagement and ineptitude, how much is due to the initial measures associated with the transition to capitalism and how much is due to the disruption of Yugoslav markets. Boarov, a Belgrade economist, wrote, "The economy was in such a catastrophic condition anyway that it's hard to separate the impact of the sanctions from bad administration."[25] Dyker and Bojicic claim that "the impact of sanctions should be seen in terms of their incremental effect on an economic trend that had already been firmly established."[26] A study of sanctions and stabilization measures conducted by the Belgrade Institute of Economic Sciences indicates that much of the downward trend began before the imposition of sanctions, and that their imposition simply solidified what was already occurring. Hinic and Bukovic point out that sanctions had a dramatic effect on the economy, with respect to both industrial production and financial and trade activity, but that the decrease in economic activity is due more to the disruption caused by the breakup of Yugoslavia than by the sanctions.[27] This is remarkable, given that Serbia was the least integrated of all the former Yugoslav regions, relying on internal markets for over 70 percent of its products. (see table 3.1). Moreover, Palairet claims that even before war broke out in Croatia, "Serbia had been deliberately reducing its trade with the northern republics in the naive belief that it could benefit from becoming more self-sufficient."[28]

The Response of the Yugoslav Government

In the period preceding the imposition of sanctions, as well as in its immediate aftermath, the government of Yugoslavia took some steps to diminish the anticipated effects of the sanctions. This "survival strategy" involved the tightening of the "collective belt," and an extension of a "collective stiff upper lip" until the international pressures subside. Even if the Yugoslav government was taken by surprise by the imposition of sanctions,[29] it did have time to make some contingency plans. Clearly, the longer the time period between the initial discussion pertaining to the imposition of sanctions and the actual imposition, the greater the possibilities of adjustment within the target state.[30] The Yugoslav government took several anticipatory steps. Some stockpiling occurred, although not nearly enough: coke and iron ore, previously imported from Eastern Europe, were in low supply even before the sanctions, and the aggravated import conditions were to have an adverse effect on the military industry. Reserves of raw materials and spare parts were not replenished as they should have been. The government also adjusted to the sanctions by developing alternative sources of supply. This included planning the movement of imports and exports through territories held by Serbs in Croatia and Bosnia-Herzegovina, which were exempt from the embargo. For this reason, soon after the imposition of sanctions, Vukovar became a thriving business center, as it was the front for trade with Serbia.[31] In fact, it is estimated that until Greece closed its northern border to all movement of oil (August 1992), some 4,000 tons of fuel a day were delivered to businesses in Macedonia and Bosnia and then resold to Serbia.[32] Furthermore, it is claimed that Hungarian and Austrian trucks carrying gasoline and chemicals frequently stopped in Pale, a suburb of Sarajevo under Serbian control. The government also attempted to stimulate domestic production, such as in pharmaceuticals, in order to decrease imports; it attempted to develop industrial substitutes, such as coal, for oil; it urged conservation of physical resources (given that there were few spare parts, and those that existed had to be conserved, the government encouraged the importation of goods, so that the stores contained imported diapers, cigarettes, coffee, detergent and shampoos—all goods that were previously domestically produced); new transport routes developed; new trade agreements and new trade markets opened, as relationships with other neighbors and friends were cultivated. For example, Serbia engaged in renewed economic relations with Romania. According to the Romanian press, "Serbia was forced to court and make advances to Romania because it had to change its previous geopolitical and economic options."[33]

The Milosevic government has printed money on numerous occasions for workers pay, pensions for the retired, price subsidies for farmers and back pay for all who received irregular payments. In short, this component of monetary policy represented an effort to pump up the economy. It was necessary in part because traditional sources of government revenue were blocked: customs duties on imports dried up, since there were no imports, and tax revenue decreased since consumption of taxable items had decreased. Other financial arrangements included the payment of high interest on bank deposits that were in fact operating with the blessing of the government. Local banks that in 1992 were paying hard-currency depositors 12 to 14 percent in monthly interest to raise money to underwrite imports thus satisfied the multitude of private depositors that used interest payments to sustain presanction living standards. The government encouraged this form of banking because it relieved pressure on itself, despite the unchecked profits that the bankers were making. Indeed, a western diplomat compared Yugoslav bankers from this period to a "cross between writers of chain letters and loan sharks."[34]

The measures discussed above were taken by the Yugoslav leadership from mid-1992 through April 30, 1993, when the tightening of sanctions occurred. New measures were deemed necessary after April 30,1993. These measures, which resembled those associated with a command economy, were described as self-defense actions and included a new list of priorities: military, energetics, food and medicines. These government sectors were to begin increased production, and those currently on forced vacations from their jobs were to be asked to participate in the production in these priority sectors. The distribution of goods in these sectors was to be taken over by state organs, since there can be no functioning market economy under sanction conditions.

On August 18, 1993, a new austerity program was introduced by the federal government. This program, like the ones preceding it, introduced greater government control over an increasing number of sectors; it introduced wage and price controls that consisted of frozen prices coupled with flexible wages, and various measures to encourage production.[35] The program received negative reviews from both members of the opposition (members of the Serbian Renewal Party said that it was an attempt to "mask past mistakes"), and members of the government (the head of Belgrade's Economic Chamber said it was too little too late).[36] However, it is unclear that under current conditions, anything other than a command economy can exist. Indeed, Kosta Mihajlovic said that "exceptional measures [were necessary] for exceptional conditions" such as those that currently prevailed.[37] Moreover, Tomislav Popovic concluded his study of sanctions by saying that after six months, the economy cannot but become a completely command economy that does not allow for stabilization measures.[38]

Sanction Busting and the Exploding Underground Economy

After the imposition of sanctions, the Serbian population is reminded of another time in history when a western power tried to bring Serbia to its knees by economic suffocation: in 1906, the Austro-Hungarian government imposed high tariffs on Serbian livestock passing through its territories (in the so-called pig war). Given that this involved nine-tenths of Serbian exports, it would have led to total economic collapse if it were not for the development of new markets and marketing routes. Similarly, today numerous ways of circumventing sanctions have emerged, both within the new Yugoslavia as well as in the bordering states. Motivated by a quick profit, people acting individually and in groups have organized themselves and are engaging in semilegal activities. On the level of individuals, activities such as the crossing of borders many times a day with tanks full of gas to be emptied inside Yugoslavia has been a favorite of Bulgarian, Romanian and Yugoslav citizens.The gasoline is sold at eight times its cost, and in foreign currency. According to a Bulgarian official, it is estimated that 300 tons of gasoline and diesel fuel were sold in July 1992 in such a manner from Bulgaria only (legislation was later passed in restricting such movement to one trip every fifteen days).[39] The town of Petric has become known as "Little Kuwait" because significant amounts of gasoline are loaded into Serbian containers there. Numerous firms have responded by increased production, since they now need not compete with foreign firms. For example, the furniture producer Simpo has been over-producing as a result of sanctions and the enlarged share of the internal market.[40]

On a more organized level, there is evidence that barges carrying oil, steel, coal and other products move into Yugoslavia by way of the Danube from Ukraine and Russia. They carry bills of landing saying their cargo is destined for Macedonia. Oil has also been said to enter through Macedonia in tanker trucks from Greece, leading Prime Minister Mitsotakis to close the northern border to all oil traffic in an effort to quell West European attacks against him. The popular press reports that chemicals are entering the country from Hungary, gasoline from Bulgaria, consumer goods from Hungary and Austria. Payments are said to have been handled through secret accounts in Cyprus. Moreover, import-export firms have been registered in Bosnia and Croatia, in cities such as Vukovar.[41] Foreign clothes, liquor and cigarettes are abundant and the supply of these items seems to be increasing with time as smugglers have time to adapt. The tightening of the sanctions on April 30, 1993, has introduced a new series of measures. For example, traffic is continuing across the border enabled by the issuance of license plates from Serbian areas of Bosnia.

However, all this smuggling activity carries a cost, which Yugoslavia cannot sustain in the long run. Indeed, a modern, export-oriented country, as Yugoslavia was before the sanctions, simply cannot revive itself on the basis of smuggling. The costs are incurred at the level of both imports and exports. It is estimated that the cost of imports has risen by some 25 percent, while the exports, when they reach their destination, have been reduced in price by some two-thirds.[42] This has been called the "sanctions surcharge."

In the short run, the stimulus that sanction-busting activity provides with respect to private initiative may have long-term beneficial effects that may offset the negative impact in the future (depending on how the international stage changes and the economic support for rebuilding after the war). This private initiative is not limited to the business of go-between that charge inflated commissions but rather extends to a wide variety of services and products. Imports of gasoline and consumer goods has become so lucrative that private enterprise, such as gasoline pumps, is emerging all over Serbia.

Political Effects of Sanctions on the New Yugoslavia

One of the goals of sanctions on Serbia and Montenegro was the toppling of the government of Slobodan Milosevic. The rationale was that the austere conditions brought about by the sanctions would so aggravate the population that they would push them to rise and rebel against "the culprit." In fact, in the Yugoslav case, as in the case of Iraq, matters did not quite proceed in the way the UN Security Council members expected.

The imposition of sanctions had a short-term effect on the political atmosphere in the new Yugoslavia. In the first place, it had the effect of strengthening the nationalist mood among the population, as sanctions were perceived as unjust and representing persecution of the Serbian people. Thus, they drew people closer to the government of Milosevic, not further away. He, in turn, capitalized on the sanctions, bolstering public morale by encouraging the population to again in history stand up to outside powers that were imposing injustice on Serbs. Thus, after three months of sanctions, it was claimed that "on balance, the sanctions may have helped Milosevic slightly by providing him with a scapegoat for the disastrous economic policies of his regime."[43] This short-run rallying behind Milosevic because of sanctions was similar to what occurred in the aftermath of Soviet sanctions on Yugoslavia in 1948-55, when the response was one of stoicism and defiance. However, this sentiment did not last. Indeed, after some 18 months of sanctions, the population is largely exhausted, weary and apathetic. Indeed, as expressed by a resident of Belgrade: "We have had enough of words; we are apathetic because we are hungry."[44] However, these sentiments do not constitute the necessary conditions for an uprising against the government.

Moreover, at the onset of the sanctions, the government clearly realized that they were going to have long-term economic and politically destabilizing effects. Hence, there was an initial receptiveness to Milan Panic, who was brought into the picture to try to persuade the international community to lift the sanctions. Personality and policy conflicts between Milosevic and Panic led to their race for president of Serbia in the elections of December 1992, in which the principal point of contention was the response to international sanctions. Indeed, the principal element of Panic's platform was a pledge to do anything necessary to ensure the lifting of sanctions and to begin economic reconstruction. From this perspective, the sanctions had one of their desired effects, namely, a domestic political confrontation. However, actions taken by the international community immediately preceding the election were so counterproductive that they raised questions as to their ulterior motive. These actions included the strengthening of the sanctions by the Security Council, which again served to strengthen the hand of the Socialist Party and Milosevic.[45] Another counterproductive action was U.S. State Department Deputy Secretary Eagleburger's accusation of Milosevic, Seselj, Karadzic and others of war crimes and his demand that they be tried. Both of these actions, which occurred prior to the elections, led to a widespread sentiment among Serbians and Montenegrians that they would not be bullied by the international community, nor ordered whom to vote for. The Balkan sentiment of *inat,* which Eagleburger is familiar with from his days at the U.S. Embassy in Belgrade, became an important factor in Milosevic's victory in December 1992.[46] The United States and the western world missed an opportunity to affect the elections in Serbia. If it had been made clear to the Serbian population that the election of Panic would have implied the removal of sanctions, the voting might have led to another outcome. Only Belgium tied the lifting of sanctions to the election of Milan Panic.

Economic and Political Effects of Sanctions on Neighboring States

The greatest possible pitfall in the success of sanctions is ensuring cooperation by the target state's allies and neighbors. In the case of the Yugoslav sanctions, the international community seems united in their efforts under the umbrella of the United Nations. However, there is clear evidence that Yugoslav neighbors did not always seriously enforce the sanctions. Greece and Russia continued to sell oil to Serbia; raw materials, and perhaps food and arms, flowed in from Romania, Bulgaria and Greece; the author personally witnessed trucks with Slovene license plates on Yugoslav highways, most probably transporting goods.[47] This persistent circumventing of the embargo continued for at least two reasons: First, the Security Council is far away, but neighbors will remain neighbors indefinitely; thus neighborly relations are severed only with caution. Second, the costs of adhering to sanctions is very high. Indeed, sanctions threaten business interests due to the disruption of trade and financial flows.

Prior to the sanctions, neighbors of Serbia and Montenegro had friendly and commercial relations with these regions, with the exception of Albania and Hungary. Relations with the latter soured after disclosure in the early 1990s that arms traffic aimed for Croatia was coming from Hungary. Romania has had the strongest links to the new Yugoslavia. Indeed, it is among the few countries, together with Russia and China, to have granted de facto recognition. Traditionally, Serbia and Romania have been friendly states, as expressed by a Romanian newspaper: "Whoever knows the history of this European region knows that Serbia is the only neighbor that has never launched an attack on us."[48] Economically, their bonds are strong: since the fall of Ceausescu and the disruption of Comecon trade links, trade with Yugoslavia has expanded greatly, 80 percent of which has gone to Serbia (total trade during 1991 grew to $440 million).[49] Part of this trade takes place in the energy sector. Imports of oil from Romania cover some 15 percent of Yugoslavia's needs, and some 20 percent, imported from Russia, comes over Romanian territory. Romania also had joint projects with Yugoslavia that would have to be disrupted at great cost. The most important of these is the hydroelectric dam on the Danube, the Iron Gates I Hydroelectric Station. There is also discussion of setting up a joint free economic zone on the border, near Timisoara, as well as some joint ventures in the Romanian Banat region.

Thus the Romanian economy has much to lose from adhering to the embargo. In June 1992, a Romanian government spokesman estimated that Romania would lose some $350 million by adhering to the Yugoslav embargo.[50] The leaders began to demand compensation. They claim Romania lost billions of dollars by joining the world embargo against Iraq in 1991 and was promised compensation that never came, and thus they will be more forceful in seeking compensation for the losses associated with the Yugoslav embargo. Without compensation, the Romanian government said "Romania itself will become a victim of the sanctions."[51]

Romania is by no means the only neighboring state affected by the Yugoslav sanctions. It is estimated by the Hungarian government that the cost to the Hungarian economy will be $300 million in 1992 alone.[52] Russia estimates its sanctions-related loss to be $16 billion.[53] Bulgarian state officials claim the country has so far lost $1.8 billion as a result of Yugoslav sanctions and expects to lose $2.5 billion annually following the tightening of the embargo.[54] It claims adherence to sanctions against Libya, Yugoslavia and Iraq will result in a total loss of $13 billion. In February, President Zhelev claimed a loss of $40 million to $60 million per month.[55] In Ukraine, the sanctions have adversely affected Danube River traffic, one of the state's main sources of hard currency. Sanctions have deprived Ukraine of revenue and caused the unemployment of 30,000 workers employed in shipping traffic.[56] Albania claims to have lost $300 million to $400 million as a result of the sanctions, and the UN Security Council agreed to send relief for Albania in September 1993.[57] Especially hurt have been Albania's chrome and electricity exports. Macedonia is in a particularly difficult position due to its adherence to sanctions. In agreeing to support the sanctions, it has severed its trade links with its northern neighbor. Its links to war-torn Bosnia-Herzegovina were severed due to the war, as were transportation routes to Slovenia and Croatia. Greece imposed its own embargo from the south as a result of disagreement over the name Macedonia is to use. Hence the closest port that Macedonia might use is closed to it. Its alternatives are the Bulgarian ports, which are 800 kilometers away, over high mountains. For these reasons, President Gligorov describes Macedonia as "sandwiched."[58] The Austrian State Institute for East and Southeast Europe has said that Eastern Europe and the Balkans will suffer losses of $356 billion in 1992 alone for sanctions.[59]

The idea of compensating neighboring states is not new nor unreasonable. Indeed, Turkey is presently requesting that the sanctions against Iraq be lifted because of their harm to its economy. Similarly, Russia has on several occasions in 1993 unsuccessfully raised the question of lifting sanctions against Yugoslavia. In the absence of such an event in the near future, compensation for the losses to neighboring states is under discussion both in the United Nations and the European Community, as well as at the level of individual governments. The German government has been quite vociferous in its claims that it wants to extend compensation. As they extended payments to Jordan, Syria and Egypt for adhering to the embargo against Iraq, so too they claim to want to help Balkan countries, including Greece, in 1992. In January 1993, when the European Community was discussing further strengthening the embargo on Yugoslavia, compensation to neighbors received top priority and was to be implemented simultaneously.[60] There has been discussion in the United Nations that various neighboring countries be awarded indirect economic compensation, such as preferential trade access to western markets, financial support for infrastructure projects or the reduction of foreign debt.[61] In addition to this Bulgarian proposal, the Ukrainian delegation to the United Nations suggested the establishment of a special compensation committee to work alongside the UN sanctions monitoring committee and to establish an emergency fund for neighboring countries affected by sanctions.[62]

SUCCESS OF SANCTIONS

The conditions for success of sanctions suggested by Hufbauer, Schott and Elliott have been applied below to the Yugoslav case in order to aid in the assessment of the sanctions in Yugoslavia.[63] According to all these conditions, the sanctions in Yugoslavia should have succeeded in achieving stated goals.

1. Companion policy, such as covert or military activity:
   As of December 1992, discussion had taken place in diplomatic circles about the desirability and feasibility of western military action in Bosnia-Herzegovina. The first step in this direction was the establishment of the no-fly zone (in October 1992) over Bosnia,

followed by Security Council approval of military enforcement of this no-fly zone. The latter act was largely a symbolic (but expensive) gesture aimed at satisfying western consciousness that "something was being done," since it was unlikely to have a significant effect on the course of the war in Bosnia.[64] Other companion policies under discussion included the lifting of the arms embargo against the Bosnian Muslims and the target bombing of the Serbian positions in Bosnia. NATO was called in to draw up plans for this military intervention, and the debate among western government officials, academics and the press took off. At the time of this writing, neither of these plans has yet been put into effect, as they have both met with strong resistance, the former from the West European leaders and the latter from within the United States.[65]

Covert action in the region has been limited to the aiding of Milosevic's opposition. This entailed the suspension of sanctions against the importation of radio and television equipment in November 1992 to enable broadcasting by the opposition into regions that Radio B, the opposition TV station, is incapable of reaching. The sort of action that took place in Iraq, when the United States sponsored the flooding of fake money into Iraq to cripple the Iraqi economy, has not yet taken place.[66]

2. The number of years economic sanctions were in force:
   The sanctions were imposed May 30, 1992, and at the time of this writing have been in effect for 18 months. However, their intensity has varied during that time, as they were progressively strengthened at various stages.
3. The extent of international cooperation:
   The Yugoslav sanctions of 1992 had the full backing of the international community. During the voting process, China was the only permanent member to abstain, while Vietnam was the only rotating member to do so. When the vote pertaining to the strengthening of the sanctions came, Serbia's traditional ally, Russia, abstained. With respect to the enforcement of sanctions, Romania, Bulgaria and Greece were willing participants in the enforcement when the United Nations called for military enforcement and the tightening of sanctions. In this respect, Yugoslavia did not enjoy the neighborly link that Iraq had with Jordan. Thus, in terms of international cooperation, there is no precedent in history for such alliance between the western powers, the former Soviet Bloc, the Muslim countries and the less developed countries.
4. The presence of international assistance to the target country:

While there is strong international backing of the sanctions, there is no offsetting international assistance to the new Yugoslavia at this time. Indeed, all regular aid, such as IMF loans, has been suspended as a result of the sanctions. When the Soviet Union imposed sanctions on Yugoslavia in 1948, international assistance was forthcoming from the West, to whom Tito turned to for military and economic help as well as trade. While there is evidence of some help from Russia in the form of goods and from Greece in the form of sanction busting, no large scale source of aid is presently available to Serbia.

5. The political stability and economic health of the target country:
If both the political system and the economic health of a target country are highly unstable and precarious, this can aid in the success of sanctions. In Yugoslavia, the economic health of the country is deteriorating day by day, as discussed in chapter 3. The political system is unstable: there are numerous political parties, but the opposition to the ruling party is not strong. While some of these parties, such as the Democratic Party, the Serbian Renewal Movement, the DEPOS coalition and others, were previously in favor of a negotiated solution to the Yugoslav crisis instead of war, the international demonization of Serbia and Serbs in general has left them without an effective platform for dealing with the evolving crisis. In October 1993, the Radical Party, under the direction of Vojislav Seselj, called for a vote of no confidence against the government of Milosevic, who survived it (by dismissing parliament and calling for new elections), despite the sanctions.
6. The warmth of prior relations between sender and target countries:
This is a difficult point to judge since the sender countries are the entire world, and the target country is a part of former Yugoslavia. With respect to the sender countries, the present divisions, including the EC, NATO, Muslim states, less-developed countries and the Eastern Bloc, all previously had friendly relations with Serbia and Montenegro as leaders of the nonaligned states. However, these relations were soured by the civil war in Bosnia, when either Croatia or the Bosnian Muslims were backed. Traditional Serbian allies, such as Greece and Russia, have continued friendly relations while reluctantly adhering to the united front that the world has posed against Serbia.
7. The costs imposed on the target country:
The effect of sanctions can more easily be measured after they have been repealed. Indeed, when Soviet sanctions against Yugoslavia ended, it was then possible to estimate their cost: $400 million

between 1948 and 1955. This figure was arrived at by calculating the Soviet credits, suspension of debt payments, the increased military budget which was deemed necessary and the welfare loss associated with the above. These costs amounted to 3.6 percent of the Yugoslav GNP in 1952.[67] While it is not yet possible to make these comprehensive calculations for the current sanctions, some efforts have been made: the Institute of Economic Sciences of Belgrade has estimated that three months of sanctions resulted in a 40 percent drop in industrial production, amounting to production of only 35 percent of the total in 1989.[68] An additional three months of sanctions would decrease that figure further to 15 to 20 percent. As this industrial production was aimed for both internal and external markets, the loss of foreign markets is estimated to be total after six months. Three months of sanctions are estimated to add 800,000 individuals to the ranks of the unemployed, and the first three months have reduced income per worker to $40. The Institute goes on to estimate that only sanctions of 1 to 1 1/2 months in duration are capable of being sustained with no cost to the economy. Any sanctions in place for longer are bound to have an escalating negative effect on the long term prospects for recovery.

8. Commercial relations between sender and target countries measured by trade between them as a percentage of target country's trade:
   That Yugoslavia is affected by sanction-induced loss of foreign trade is evident: according to Milojcic, 8.5 percent and 8.9 percent of Serbia's and Montenegro's production is aimed at external markets, and imports amount to 8.7 and 10.5 percent, respectively, of its GNP.[69]
9. The relative economic size of the countries (sender and target):
   In this case, the new Yugoslavia is pitched against the entire international community. The new Yugoslav federation, compared to the sender countries, is an insignificant speck with respect to population, territory and GNP.
10. The type of sanctions imposed:
   The sanctions imposed on the new Yugoslavia are the most comprehensive blockade of a state and its peoples that has ever been achieved in modern history. Isolated examples such as the siege of Masada stand out as historical analogies but are of little use in the modern world. There is little left to do to further isolate Yugoslavia other than placing armies at its borders, forbidding the private flow of individuals across borders and cutting off telecommunications and telephone services. As it is, the former Yugoslav president Dobrica

Cosic has remarked that the current sanctions have created Serbian reservations in the middle of Europe and that an entire nation is held hostage.

11. The cost to the sender country:
 Unequivocally, the United States is the principle sender of the sanctions. It is the U.S. government that pushed for their imposition in mid-1992, and for their strengthening in mid-1993. The cost of the sanctions to the United States in the form of lost trade is negligible, while the cost of enforcements, such as patrols of the border, is only slightly higher, since patrols are composed of largely European personnel. However, the cost to neighboring Yugoslav states that support the sanctions are large. These estimates were enumerated in the text above; suffice it to say that they are strangling the economies of neighboring states.

According to the variables that affect success of sanctions enumerated above, the success of sanctions in the new Yugoslavia should be high. But an assessment of success entails a renewed discussion of the goals. Insofar as the goal was to topple and destabilize the Milosevic government, thus far the sanctions seem to have failed. In fact, their element of bullying has introduced an anti-western sentiment that has served only to strengthen the resolve of the Serbs, and thus to rally them around President Milosevic. Even the political instability that the sanctions were to have caused has been minor and insufficient to cause real change. The goal of forcing agreement to the Vance-Owen Peace Plan was achieved, since sanctions have contributed to the change of heart of Milosevic and Cosic, as expressed in their letter written to the Bosnian-Serbian parliament in May 1993, prior to that body's rejection of the plan.[70] Thus while sanctions may have contributed to Serbia's pressure on the Bosnian Serbs, it is unlikely that they will have a strong impact on peace on the ground, at least not in the short run. Indeed, the fierce fighting between Muslims and Croats in central Bosnia at the time of this writing is a further indication why the sanctions on Serbia are likely to have little effect on peace on the ground since such a peace involves other parties.

CONCLUSIONS

As noted above, sanctions vary in their intensity and comprehensiveness. Those sanctions that have not been comprehensive have also not succeeded in achieving their goals. Indeed, Cuba is still ruled by Castro, despite three decades of sanctions; South Africa has only recently changed its political landscape, although certainly not as a result of sanctions; the Haitian generals are still in power at the time of this writing, while deposed President Aristide wrings his hands in New York; and Gaddafi has not relinquished the suspects in the Pan Am bombing. Yugoslavia and Iraq stand out from the others in the comprehensiveness of the sanctions against them: not only do they include a wide range of measures, but they enjoy virtually universal support from the world community. In their comprehensiveness, they parallel the Middle Ages custom of papal interdiction on an entire country in order to alter behavior. By all indications, Yugoslav sanctions had the greatest chances of success.[71] Despite these indications, at the time of this writing, both President Milosevic and Saddam Hussein are still in power, and there is no indication that their governments are wobbling. The most significant result of the sanctions has been in the humanitarian sphere. In both states, there has been a dramatic decrease in the standard of living of the population, as basic needs such as health care and food accessibility are no longer satisfied. Measures that so utterly impoverish a people are simply not conducive to producing an uprising against the ruling structure. (In fact, instead of weakening the government structure, they have weakened the opposition. Sanctions have had the effect of pushing an embattered and embittered civilian population to rally around the very power structure that is the butt of sanctions.) Apathy and devastation, along with regional instability brought about by a lack of economic development, are the more likely consequences. Indeed, the goals of sanctions are contrary to their effects: the goal is to bring about a democratic government that is in the interest of the population, but the effects are the infliction of pain on that same population. Is inflicting widespread hardship acceptable, even if in the name of democracy? The sooner the U.S. government, the chief instigator of sanctions in the post–cold war world, recognizes that fact, the less humanitarian damage will be suffered by numerous innocent peoples across the globe.

In addition to the humanitarian aspect of sanctions, there are two other issues that warrant discussion in the case of Yugoslav sanctions: the questionable logic of applying sanctions to Serbia for a war in Bosnia and rebuilding in the aftermath of sanctions. With respect to the first point, one needs to be reminded that one of the goals of those sanctions was to bring about peace in Bosnia-Herzegovina by applying pressure on Serbs in Serbia. The logic behind this is faulty for several reasons. First, it is simply not clear what the participation of Serbs in Serbia has been during the war. While support definitely existed, it is unclear how strong it was and how much independence the Bosnian Serbs actually have. It increasingly seems that they have much. Second, there is no logic in applying sanctions to one indirect backer of a side in a civil war and not to others. There is clear evidence that Croatia has been aiding the Bosnian Croats in many ways. Indeed, Croatian troops are known to be actively participating in the ground war in Herzegovina. Thus it seems that if sanctions are to be applied to stop the war, they must be applied to all participants.

With respect to the aftermath of sanctions, whenever they may be lifted, it is important to note the degree of the devastation of the Yugoslav economy. It is now clear that Yugoslavia cannot hope to recover in the twentieth century. One day the recovery will begin, however, and it will have to begin with an infusion of capital from outside its borders, although from where is not immediately clear. While the rebuilding of Croatia will be aided by both Germany and international organizations that have been supportive in the course of the past two years of warring and the reconstruction of the Muslim regions will be equally quick given the commitments made by Muslim countries (most recently, in April 1993, of $90 million), the reconstruction of Serbia is going to be harder to fund. Zaslavsky reminds us of the experience of defeated Germany and Japan, both of which had highly militarized and monopolistic economies, dominated by a strong government.[72] They made remarkable recoveries, but those recoveries were only possible because of the destruction of economic and political structures that had occurred, decades of hard work on the part of their populations, which understood that it was necessary for survival, foreign investment, foreign troops to ensure stability and probably also a culture generally predisposed to hard work. Could such a destruction of Serbia lay the groundwork for a similar reemergence? It is unclear at this point, especially given the fact that the international community would not look favorably at an inflow of funds, especially under the present government. This sentiment is exemplified by rumored demands that Serbia be forced to pay war reparations. While the record of war reparations in the Balkans is varied (Romania never paid reparations after World War II, Croatia still needs to pay to Italy, while Greece paid heftily to Turkey for its invasion in 1922), reparations are often demanded after wars. Presently several nonaligned countries are calling for the extraction of reparations from Serbia for the destruction of Bosnia-Herzegovina, while the U.S. ambassador to Croatia, Peter Galbraith, is in favor of tying the lifting of sanctions to the payment of reparations for damages inflicted on Croatia.[73] Both of these, while faulty in their analysis for all the reasons mentioned above, are also faulty on economic grounds. As pointed out by John Maynard Keynes in the aftermath of World War I, it is futile to try to extract reparations from a vanquished power that does not have the capacity to pay. While Keynes was referring to Germany and the Allies' attempt to extract reparations, the argument applies to the new Yugoslavia also, since even if reparations were deemed justified, would not have the capacity to pay. Dyker and Bojicic point out that the West will instead have to participate in the reconstruction of Yugoslavia, and it would be ludicrous "to have these resources commandeered by other countries under the title of reparations."[74]

Given that the sanctions imposed on Serbia and Montenegro have not achieved their stated goals but have caused many deaths due to deteriorating health conditions, cold and hunger, their continuation leads one to question the goals of the international community. If sanctions did not end the war in Bosnia (note the intensification of fighting between Muslims and Croats in central Bosnia), they did not cause a retreat by Bosnian Serbs from the territory they control, nor did they topple the Milosevic government, yet they persist despite humanitarian warnings by WHO, then one is led to believe that there might be underlying motives that have not been spelled out by the UN resolutions. This question is increasingly being put among academics, policy makers and the media. Supporters of both the current rulers in Yugoslavia and of the opposition have raised questions as to the underlying reason for the Yugoslav sanctions. For example, Jarcevic, the chief of diplomacy of the Serbian Republic of Krajina, has suggested that the sanctions might have an economic motive, such as the prevention of the new Yugoslavia from competing with western products in the Eastern Bloc countries, where their relatively low prices would have made them competitive.[75] In addition, according to *Vreme* (the Belgrade opposition magazine), "The blockade did not blow away the top leadership from the political scene or stop the war, but it can be proved that, *in the strategic sense* [italics added], it has disabled Yugoslavia from restoring a normal pace of development until the end of the twentieth century,"[76] implying that the destruction of the Yugoslav economy was at least one of the ulterior motives of the sanctions. While it is dangerous to hypothesize about motivation, and naive to give international bodies extensive credit for their clear thinking and foresight, there is an appeal of the argument that indeed the destruction of the Serbian economy might be, if not a stated motive, at least a welcome by-product. As evidence in earlier chapters suggests, the Serbs are among the largest ethnic groups in the Balkans; their territory, together with Montenegro, covers a wide area of the Balkans. If one day the Serbian-populated territories of Bosnia and Croatia were added, then this new Yugoslavia would cover a substantial chunk of Balkan territory.

Not only would this Serbian-dominated Yugoslavia then be a large economic unit, with abundant human and natural resources and extensive markets, but it would also be a major Balkan power. This has been recognized by members of the German government, who, in an interview in the *Observer,* noted that Serbia has become the "regional superpower."[77] Indeed, one must not forget that the new Yugoslavia inherited most of the holdings of the Yugoslav army. Former Yugoslavia had the fourth largest army in Europe, and much of that strength has been retained by the current Yugoslav army. Since the international community cannot control Yugoslavia's purchase and production of weapons directly, it can do so indirectly by controlling the economic infrastructure that supports the weapon industry. Thus, sanctions have the geostrategic effect of controlling the military buildup and the creation of a military Balkan power by economic means. The sanctions may thus have, as a byproduct, the effect of altering the relative power structure in the Balkans.

6

The Economic and Political Ramifications of Population Movements

Population movements have been a persistent component of world history. People have left their homes in pursuit of better jobs and political or religious freedoms. Some have fled from persecution, wars or natural disasters. Indeed, both push and pull factors contribute to the migrant's decision to relocate. Migration affects not only the lives of the migrants, but also affects the gaining and losing regions. Large waves of economic migrants have enabled economic development in host regions, including the United States, the Middle East and Germany. Losing states have suffered in numerous ways, including as a result of brain drain, such as that associated with the Jewish emigration from the Soviet Union and the politically based exodus from pre-World War II Germany.

Since the turn of this century there has been an increase in the number of politically induced population movements. The communist revolution in Russia produced some 1.5 million refugees, while Turkish policies induced the movement of some 250,000 Armenians and later over one million Greeks. During and after World War II, Hitler's government induced migrations of ten million people in Eastern Europe, while the partition of India caused a displacement of over ten million Muslims and Hindus. More recently, the creation of Bangladesh uprooted over ten million people, Sudan accepted approximately 350,000 Eritreans, Somalia took in 800,000 people fleeing Ogaden Province, the Soviet invasion of Afghanistan sent two million people into Pakistan, Iraq's invasion of Kuwait sent 380,000 Palestinians out of the Persian Gulf, 80,000 Cubans fled Cuba over the course of a few weeks, and 100,000 Jews fled the Soviet Union in a few years during the 1970s. From 1979 to 1989, some eight million people were driven from their homes by "superpower proxy wars" in places such as Afghanistan,

Cambodia and El Salvador.[1] In the first three years after the end of the cold war (1989-92), some 4.5 million refugees were produced as people fled interethnic strife. In Africa alone, rampant interethnic wars have resulted in a situation in which only four African states have neither gained nor lost over 1,000 refugees.[2]

The present refugee problem associated with the civil war in former Yugoslavia has received much notoriety in 1992-93. It is estimated that there are some two million displaced persons in the territory of former Yugoslavia, and the number is growing daily. Indeed, the U.S. State Department estimated that if Bosnia becomes partitioned into three ethnic states, it would result in the relocation of 600,000 more Muslims, some 300,000 Croats and 350,000 Serbs.[3] The latest threat of the Yugoslav government to deport some 500,000 Serbian refugees back into war-torn Bosnia increases the horror,[4] as does the Croatian threat to withdraw refugee status of 250,000 Bosnian Muslims on their territory, in the wake of conflicts between Bosnian Croats and Bosnian Muslims in Mostar (May 1993).[5] While these numbers pale in comparison to the number of Afghani or Palestinian refugees (4.6 million and 2.7 million respectively), it nevertheless represents a significant demographic shift of a prewar population of 23 million. The economies and societies of Croatia and Serbia, the host states where the largest number of refugees have been housed, are bursting at the seams. Large numbers of refugees from former Yugoslavia are dispersed throughout Western Europe, where governments are claiming they can house no more. Indeed, their present reluctance is understandable given the saturation with refugees that is manifested in antisocial actions by the dissatisfied and unemployed youth (for example, in Germany and Italy). Indeed, the refugee-weariness of the 1980s is turning into refugeeophobia of the 1990s, as in-migration becomes a political issue and the ills of western states become increasingly blamed on the incoming migrants (witness the platform of the Lombard League in Italy). It must be noted, however, that the reluctance is largely limited to the West, in part because it is the most desired destination: considering the receptive welcome refugees have received elsewhere in the world in the past decade, such as Malawi (which is hosting 950,000 refugees although its population is only 9.5 million people) and Pakistan (which has taken in 3.6 million Afghan refugees), the western effort pales in comparison.

Population movements, whether voluntary or involuntary, carry with them economic, political and demographic implications not only for the migrating individuals but also for the host and losing regions. Migration represents an opportunity, and carries a cost, at both the level of the state and the individual. These benefits and costs are studied in this chapter. Population movements in modern Balkan history, especially those caused by the disintegration of Yugoslavia, are studied from the point of view of nationalist bankruptcy. Indeed, these population movements are simultaneously concomitants of nationalist bankruptcy as well as one of its causes.

INVOLUNTARY MIGRATIONS

Population movements can be divided into those that are forced and those that are voluntary. Some movements are clearly one or the other, such as the forced resettlement of Cossacks during the Stalin era and the economically motivated immigration from Italy to America at the turn of the century. This distinction between what is voluntary and involuntary migration has become especially poignant in the recent revival of "ethnic cleansing" in the former Yugoslavia.

Forced, or involuntary, migration, produces refugees. Refugees have been defined as those individuals that flee from man-made disasters. Gordenker defines refugees as "persons who have left their customary homes under the pressure of fear for their present or future lives, because of immediate, overt threats or—more comprehensively—clear denials of basic human rights whose enjoyment is required for continued life over a short or longer period."[6] Refugees flee from a set of international or domestic circumstances. Among the former are wars involving armed intervention and/or political warfare, such as propaganda or a victorious new political system. War situations provoked population movements in Indo-China following the victory of North Vietnam and Jewish emigration from Germany, Poland and Croatia during World War II; Greeks and Turks exchanged residences following the Greco-Turkish war of 1922. New political systems were responsible for the pressure on Germans to leave the Soviet Union and Eastern Europe following World War II and Asians to leave Uganda in the 1970s. Wars of liberation or decolonization provoked a mass exodus of Portuguese from Mozambique and whites from Malawi and Rhodesia. Another component of the international circumstances of refugee creation is the redrawing of borders in peace, when it puts national, religious or racial groups on the wrong side of a border. This then causes mass population movements, such as the population movement of Russians after the breakup of the Soviet Union as well as the movements of various Yugoslav ethnic groups in the aftermath of the Yugoslav breakup.

With respect to domestic pressures on populations to relocate, the most important is turbulence of various forms: a violent governmental change, such as one associated with a revolution or a coup d'etat that carries with it either policies adverse to a given people or simply violence, creates refugees.

For example, the revolution in Nicaragua in 1979 caused significant population movement for both of these reasons. Further, persecution on the basis of religion, race or ethnicity, whether by sporadic harassment or planned genocide, results in refugees (for example, the Armenian exodus from Turkey, the Indian flood from Guatemala, the Muslim exodus from Myanmar). However, minorities are not the only ones suffering from what might be a brutal, dictatorial government that harasses members of society: the political opposition may also be a target, resulting in political migration (for example, the exodus of political opponents of Pinochet from Chile or of the Ayatola Khomeini from Iran). Not all governments that cause population movements do so out of design or malice. Indeed, some are simply incompetent, and are unable to lead their populations or offer them adequate standards of living. For example: government incompetence led to hunger in Somalia and Ethiopia, while political instability in Lebanon in the 1970s led to chaos, and both created destabilizing population movements.

Given that involuntary migrations have been present throughout history, why are refugees receiving so much attention in the 1990s? Gordenker claims that the post-World War II involuntary migrations have several novel characteristics.[7] First, the sheer volume of refugees has risen dramatically since 1945, as has information of their plight through the mass media. Second, most refugee situations since the war have taken place in less-developed countries, with the exception of Hungary in 1956 and former Yugoslavia in 1992. Such movements of population to and from developing countries have influenced the development plans of both the losing and the receiving region. Third, many of the population movements have been on such a large scale and have remained unresolved for so long that they have become permanent, putting strain on the immigration system of the receiving state. Indeed, the Palestinians from Israeli territory and the Chinese from Vietnam are examples of generations living outside of their native territory. Fourth, there has been an unprecedented increase in international organizations to assist refugees, in response to the sheer volume of displacement. Although most of these have been through the UN systems, there are also numerous efforts at the bilateral level.

POPULATION MOVEMENTS IN THE BALKANS

Since the end of World War I, there have been over a dozen population movements in the Balkans that involved over 20,000 people.[8] While some have been clearly voluntary or involuntary, the vast majority have been in the shady region in between. While there exists documentation on large scale movements of ethnic and religious groups, voluntary movements are rarely classified by ethnic group, since ethnic persecution is not the critical issue in the decision to migrate.

From 1919 to 1924, 80,000 Hungarians moved from the former Yugoslav territories to Hungary and 20,000 Turks moved to Turkey. During this period, Romania lost 200,000 Hungarians to Hungary, 80,000 Turks to Turkey (1913-39), and 50,000 Bulgarians to Bulgaria.[9] From Macedonia, 30,000 Bulgarians moved to Bulgaria, as did 70,000 from Turkey and 120,000 from Greece. Bulgaria also lost 110,000 Turks from 1913 to 1939. Also, 350,000 Turks left Greece for Turkey, while 1.3 million Greeks left Turkey for Greece during 1922-24. When Albania was created and its present borders fixed, Greeks emigrated to Greece. These population movements in the post-World War I period do not include the inflow of people from non-Balkan states, such as that of the Russians following the Bolshevik revolution or the Armenians fleeing persecution in Turkey.

World War II again witnessed massive migrations across the Balkans. Most of the movements pale in comparison to the migration of Germans from Poland (over eight million) and former Czechoslovakia (over 2.5 million). Germans were also forced to leave the Balkans, while Italians and Serbs departed from Croatia. It is estimated that some 100,000 Serbs left Kosovo during the Albanian occupation, none of whon returned from their safe haven in Vojvodina.[10] Moreover, between the late 1960s and mid-1980s, between 200,000 and 300,000 more Serbs left Kosovo.[11] Hungarians left Romania after World War II, as did Jews who survived the war.

Since World War II, there have been continued outflows of Jews from all Balkan regions. This exodus received notoriety in Romania, where the policy of the Ceausescu government was to sell exit visas for $10,000 per head. Another population movement that has received attention is that of the (forced) exodus of Turks from Bulgaria in the mid to late 1980s, when 160,000 Turks may have entered Turkey. [12]

The turmoil in the former Soviet Bloc gave rise to large population movements. By 1991, after four years of continuous inflows, Hungary received 50,000 refugees, of whom 85 percent were from Romania, and 80 percent of those were Hungarians.[13] At the same time, Greece and Italy were recipients of a wave of Albanian emigration. In Greece, it is estimated that 200,000 Albanians have entered the country since 1990. The Greek government countered this inflow with the policy of the "Iron Broom" (started in December 1991), which succeeded in repatriating 1,300 Albanian refugees back to Albania in a single week in February 1992 on the grounds that they were economic rather than political immigrants.[14] Italy returned thousands of people in boats that tried to land at the port of Bari. Oddly, parts of Yugoslavia were also recipients of refugees from Albania: over 1,000 registered in Montenegro in January 1992, and 300 in Macedonia in March 1991.[15] Bulgaria has been the recipient of ethnic Turks wanting to return to Bulgaria, some 200,000 having already returned in 1991. Adding to potential problems are also the estimated 700,000 Bulgarians in the former Soviet Union that want to return to their homeland.[16] Romania is the home state of hundreds of Gypsies that the German government has deported and is reportedly paying the Romanian government to repatriate. It is also reported to have promised to pay for the repatriation of 25,000 Bulgarians.[17] Moreover, Romania lost 100,000 ethnic Germans in 1990 alone.[18] In addition to receiving refugees, Balkan countries are also losing population. Outside of former Yugoslavia, this is most evident in Bulgaria and Albania. In the former, the estimate of the loss to emigration ranges from 138,000 to 400,000, mostly young people.

Undoubtedly, refugees from former Yugoslavia have received the greatest attention in the 1990s, although according to the numbers, they are by no means the most significant group in size (compared to refugees from Angola, for example). The Yugoslav war has produced both internal and external refugees. With respect to the internal refugees, the largest concentration is presently found in Serbia and Croatia. It is estimated that by the end of 1992, there were two million refugees, and three million displaced people were receiving assistance.[19] According to Croat statistics released in January 1993, since 1991, there have been 515,000 refugees, plus 285,000 registered from Bosnia-Herzegovina and 85,000 nonregistered people.[20] Some 540,000 refugees from Bosnia-Herzegovina passed through Croatia to other destinations, and 55,000 Croats have taken refuge outside the territories of former Yugoslavia. By November 1992, the Serbian commissioner for refugees claimed that there were 545,000 registered refugees and estimated that there were 120,000 additional nonregistered ones.[21] In early 1993, the Serbian Red Cross released figures pertaining to the 600,000 registered refugees, indicating that more than half are from Bosnia-Herzegovina, 217,000 from Croatia, 37,000 from Slovenia and 3,000 from Macedonia.[22] Of these refugees, only 84 percent are Serbs, and 215,000 are under age 18 (and half of those are under age 7). By the end of 1992, some 630,000 refugees were in Croatia.[23] Two new efforts have taken place to attempt to curb the refugee problem in Croatia and Serbia. First, in June 1993 efforts were under way to expel the 250,000 Muslim refugees that are housed, at government expense, throughout Croatia. A second, new regulation on April 25, 1993, disqualified Serb refugees from Serb-held territories in Krajina and Bosnia-Herzegovina from applying for refugee status in Serbia.[24]

Table 6.1
Refugees from Former Yugoslavia in Europe, August 1992

Country	Number of Refugees
Turkey	15,000
Greece	7
Hungary	50,000
Czechoslovakia	4,000
Poland	1,500
Austria	57,000
Italy	17,000
Switzerland	70,450
Luxemburg	1,200
Germany	220,000
Denmark	1,795
Britain	2,000
Ireland	10
Netherlands	6,300
Belgium	1,800
France	1,108
Spain	120
Norway	2,617
Sweden	47,600
Finland	1,892

Source: *La Repubblica,* September 24, 1992.

Refugees from former Yugoslavia have also found their way into neighboring countries. The major recipient country was Austria, which, according to table 6.1, is housing 57,500 refugees. Austria is followed by Hungary, which had taken in 50,000 refugees (mostly Croats) by the end of 1991. Ethnic Hungarians, refugees both from Croatia and Vojvodina, have also emigrated to Hungary.

ECONOMIC SIGNIFICANCE OF POPULATION MOVEMENTS

Individuals

In different ways, both the voluntary and involuntary immigrant makes a cost-benefit analysis of moving and acts according to its outcome. While the odds for involuntary immigrants are largely against their staying (as in the case of Indians in Uganda, or more recently, Muslims in eastern Bosnia or Serbs in western Herzegovina), they nevertheless make a rational assessment, which may factor in death as a cost.

The present outflow of emigrants from the Balkans, especially from former Yugoslavia, is composed of both refugees that are victims of various forms of ethnic cleansing and those that are economic emigrants acting on their own free will in the wave of refugees. However, all of them have met with a measurable lack of acceptance in the western states. Indeed, we seem to have reached what Mesic described as the "crisis" stage of migratory flows, namely the one in which the exit is free but the entry into other countries is closed.[25] The flow of neither economic emigrants (*gastarbeiters*), nor political refugees is presently encouraged.[26] Indeed, in the spring of 1993, the Croatian authorities were turning back busloads of Muslim refugees that had no certification of either transit visas or family support. Germany has gone so far as to sign an agreement with the Romanian government to repatriate Romas (Gypsies): it is estimated that $20 million is being given to reintegrate some 50,000 total Romanian Gypsies.[27] Germany is not alone in asking refugees to leave: there is evidence from Iran that Afghan refugees are being sent home and that authorities are requesting western assistance to complete the process.[28]

Receiving Regions

In addition to the impact on the individual, migration makes an economic impact also at the level of the state. The receiving region makes a cost-benefit analysis in advance of opening its doors to immigrants. The greatest potential benefit to the receiving region is that derived from the inflow of skilled workers whose training took place at the expense of another government—in other words, "brain gain" which complements another country's brain drain. Indeed, countries such as the United States and Switzerland have been absorbers of qualified workers for decades: it is estimated that the United

States saved some $4 billion in training costs of its labor force from World War II until the 1960s.[29] The benefits of skilled immigrants extend beyond their earning and productive capacity to the general atmosphere they bring with them: for example, the Cuban community in south Florida has invigorated the area with its enthusiasm and drive.

Another potential benefit to the receiving region is the inflow of money to help manage the new population. This ranges from the financial input that regions such as Croatia have received from the German and U.S. governments in the course of 1992 to help cope with the refugees from Bosnia to the direct payment by the German government to the Romanian government to take back refugees that have relocated to Germany over the course of decades. These direct payments can motivate a country to absorb both involuntary and voluntary immigrants. The UN High Commissioner of Refugees estimated that the budget necessary to deal with displaced peoples from the Yugoslav war is on the order of $15 million per month.[30] In February 1992, the International Red Cross announced that it would have to more than double its budget from 16 to 37 million Swiss francs to deal with displaced peoples in the former Yugoslavia, and the Croatian government announced that it was spending 3 million deutsche marks per day caring for 320,000 refugees.[31] The United States has contributing $6 million to refugee assistance from former Yugoslavia.[32] In the Balkan crisis today, the receiving regions are able to reap benefits in the form of financial help for the refugees.[33]

In addition to the skill of its immigrant workers, receiving regions have the benefit of a population willing to perform undesirable and dangerous jobs, many of which would often not be performed by indigenous workers (witness migrant farm workers from Mexico in the United States or street sweepers from Turkey in Switzerland). Furthermore, these workers are often not adequately protected by laws and thus may be easily dismissed in the case of an economic downturn.

Costs are incurred by the receiving region. The most direct costs are those associated with resettlement, even when it is temporary: policing, transportation, water and health controls, and so on. However, there are other costs that are indirect and often take time to emerge. Most notable among these is the jealousy of the population that perceives that its livelihood is threatened, its costs are rising and the competition for scarce resources has been unfairly sharpened. These issues have spurred a rise in anti-foreign sentiment across Germany in the early 1990s.

When the result of the cost-benefit analysis is negative, the costs of accepting refugees leaves a mark on the state economy. The magnitude of the pain associated with refugees is perhaps best exemplified by the effects on the Greek economy after the population exchange of two million people with Turkey in 1922-23: Mazower attributes some aspects of the Greek slump in the late 1920s directly to the refugee crisis.[34] More recently, the economic crisis in Yugoslavia can in part be explained by the increased demand and shrinking supply of goods and services caused by the influx of refugees. The governments of both Serbia and Croatia are claiming that the costs associated with the inflow of refugees is presently too large for their societies to bear, a sentiment that is manifesting itself in the desire of both Presidents Milosevic and Tudjman to speed up the end of the Bosnian war. This strain is real, despite the fact that, most refugees are currently housed by family and friends: indeed, relief officials say that Serbia has lodged 96 percent of its refugees in private homes, a situation almost without precedent.[35]

Losing Regions

The most pronounced loss associated with the outflow of emigrants is in terms of human capital (brain drain). While scholars disagree as to how to measure the magnitude of the loss,[36] they agree that the price a losing region pays in terms of economic growth potential is great. Losses are incurred when a state pays for education and training of individuals that another state then gets cost-free. Clearly, the more trained the emigrants, the greater the loss. From ancient times, brain drain has been a problem to societies: indeed, the movement of intellectuals from Athens to Ptolemy's library in Alexandria and from Constantinople to Western Europe led to the awakening of one culture at the expense of the other. According to Gustave Arlt, "Without the brain drain from Constantinople [1204-1453], it is hard to imagine what the later history of Europe might have been."[37] More recently, a large movement of human capital left Europe for the United States to escape conditions associated with both national socialism and communism. In the late 1980s and early 1990s, there has been much population movement out of the Balkans. It is estimated that 20,000 educated Albanians left Albania between July 1990 and February 1991 in search of employment opportunities.[38] Bulgaria also lost some 200,000 people in 1989, most of whom were young people: this is expected, according to Joly, to have a very damaging effect of the economic development of the country.[39] Official Bulgarian figures claim that in 1990 alone, 248,000 people left the country, and over 60 percent were between the ages of 15 and 39.[40] From Serbia and Montenegro, there is evidence that 100,000 to 150,000 educated professionals left during 1992, leaving open numerous positions for academics, engineers and doctors.[41] The replacement cost to the Balkan states is high. It has been estimated that each B.A. takes some four years to replace, while Ph.D.s take seven years.

In addition to brain drain, the losing region suffers other smaller costs when the emigrant departs. These values depend on the emigrant and the circumstances: loss of the emigrant's savings (resulting in a decrease in the rate of investment), loss of the emigrant's fertility (resulting in a decrease in the future human capital pool), loss of the emigrant's taxes (resulting in a decrease in government revenue), loss of an intangible contribution to the labor force, such as leadership (resulting in a decrease in productivity of others), and so on.

THE DEMOGRAPHIC STRUGGLE FOR POLITICAL POWER

The economic and political implications of population size are not to be undervalued.[42] The sheer size of a population is associated with might, as is evident in the cases of China and India, both of which are important regional powers. Throughout history, the drive to increase population has led to conquests, receptive immigration laws and pronatalist population policies. But population size has taken on a new meaning in the post-cold war world, in which nationalism is on the rise and ethnic self-awareness seems to permeate once placid populations. Presently, the size of a total population is less relevant than the absolute and relative size of a particular ethnic or religious group. This interest in the numbers of ethnic groups is reflected in the renewed interest in statistical recordings that count births, deaths, residences and migrations by ethnicity. Indeed, the Bulgarian census is awaited with trepidation, as it is the first compilation of data on ethnic affiliation in over 15 years;[43] the Indians play guessing games as they try to predict the strength of the unruly populations of Kashmir, Assam and Punjab between population censuses; the authorities in Myanmar try to "rectify" the religious population balance before census workers go into the field by encouraging the out-migration of Muslims from Arakan Province. It is because of the political significance of population size that the Albanian populations of both Kosovo and Macedonia have boycotted recent censuses, enabling them to claim a larger number of people than the censuses might indicate (the official census figures claim that Albanians make up 21 percent of Macedonia's population, while leaders of the Albania community insist that the figure is closed to 40 percent).[44] The power of numbers is also clear in the current civil war in Bosnia, where all three sides are using population statistics to bolster claims that their populations deserve more territory and political power. According to Bogosavljevic, "Members of many nations in the territory of former Yugoslavia are vying with stories on who has not been properly counted, their numbers, who 'dominates' and who has the right to more sunshine."[45]

Thus, in the present climate, in which ethnicity is important, how does a government that wishes to alter the composition of its population go about achieving such a goal? Possible measures include policies pertaining to population growth, mass conversions, ethnic dilutions (including ethnic cleansing), immigration and territorial expansion. These are discussed below.

1. A pronatalist population policy entails the stimulation of procreation in order to increase fertility rates and, over the long run, increase population. A passive pronatalist policy may simply entail urging people to procreate (such as the call on Jews worldwide by their religious leaders in response to falling birth rates), while an active policy may entail direct monetary compensation (as in Italy under Mussolini), financial stimuli (as in Singapore during the 1970s), prohibition of birth control (as in Romania under Ceausescu), et cetera. In the Balkans, a passive pronatalist policy has been explicit for decades among the Albanians of Kosovo and Macedonia. The encouragement of large families by both religious and political leaders is partially responsible for creating the highest population growth rates in Europe: indeed, Albanians in Kosovo are increasing at a rate of 2.5 percent per year (while Croatians are increasing by 0.4 percent). According to Flere, the birthrate in Kosovo is calculated to be 30.4 (per 1,000), compared to 15.3 in all of Yugoslavia.[46] In the course of the ethnic awakening of the early 1990s, religious leaders in both the Catholic and Orthodox churches have called for increased populations of Croats and Serbs respectively. The Balkans have also had active pronatalist policies. While all former communist states had generous social policies that included maternity leave and child-care facilities, the Romanian government went the farthest, prohibiting birth control and illegalizing abortions. While these policies were universal, there were "ethnic loopholes" since they were not intended to encourage the birth of, for example, Gypsy babies.

   Waiting for pronatalist policies to alter fertility rates and to translate into increased population requires patience that many governments do not have in a climate of competing nationalisms. Since time is crucial, other measures, described below, are more rewarding insofar as the results appear faster.

2. The second measure that governments take entails mass conversions. This applies to situations in which competing populations differ in religion or language. As in the case of pronatalist policies, mass conversion may be passive or active. In the former, populations are encouraged, cajoled and stimulated to convert. The prize may be acceptance into the dominant culture or financial gain (for example, the Ottoman Empire in its Balkan dominions was known to have compensated and rewarded those that converted to Islam, notably the Muslims of Bosnia or the Albanians). Active conversion attempts are common: the Bulgarian government in the 1980s engaged in massive Bulgarization of the Turkish population by forcing name changes, while the Russians forced the Cyrillic script on numerous Arabic, Turkik, and Moldavian populations.

   Mass conversions may be perceived as giving a population a chance to make themselves acceptable to the rulers, in exchange for the right to live in their present location. Failing to convert often carried with it dire consequences, for which the rulers exonerated themselves because they blamed the population for the consequences of its refusal to convert. Usually, the consequences of refusal to partake in mass conversions (or when the indigenous peoples were of an ethnic stock too different and too difficult to convert), were expulsion or murder of the undesirables. That the Bulgarian government, in the 1980s, simply expelled a large number of Turks who were resisting conversion may be viewed, with the hindsight of the Yugoslav war, as a tame measure. Indeed, the Yugoslav territories have dealt with the issue of conversion in the civil war of the 1940s as well as the 1990s. In the former, the policy of the Ustasha government was, as stated by the education minister at a banquet, the following: "One-third of the Serbs we shall kill, another we shall deport, and the last we shall force to embrace the Roman Catholic religion and thus meld them into Croats."[47] In the latter, Serbian, Croat and Bosnian Muslim nationalists have engaged in forced expulsion of enemy populations in Bosnia-Herzegovina in an effort to readjust the regional ethnic composition. While the Serbian forces committed the largest number of expulsions, no side has been blameless in their efforts to ensure ethnic purity.

3. Governments may affect the ethnic composition of a population by systematic dilution. This process entails both a push and a pull. An

ethnic group may be diluted because members of a different ethnic group are pushed onto their territory. This type of resettlement, while it may resemble others practiced across the globe, differs from those with the goal, for example, of increasing the labor supply (such as in Siberia) or alleviating population pressures (such as in Malaysia). Indeed, ethnic resettlements target individuals not by their skills or income levels but rather by ethnicity (or religion or language), with the goal of diminishing the political and cultural strength that the group derives from its numbers. Examples of dilution by push include the dilution of the Golan Heights and the Gaza strip by the infusion of Israeli citizens into these occupied territories: the dilution of the Tibet due to the forced in-migrations of the Han; the dilution of western Poland by Hitler through the importation of Germans from across Eastern Europe; and in the Balkans, the dilution of the Hungarian population by the resettlement of Romanians into Transylvania and the dilution of the Istrians by the relocation of Croat refugees from Slavonia, Krajina and Bosnia (Joly points out that the Croatian refugees were relocated to Istria "in order to dilute the Italian-speaking population there and possibly bolster support for the ruling Croatian Democratic Community party in an area where they had previously been defeated in elections").[48] Serbian refugees are being resettled in eastern Slavonia from other parts of Croatia: indeed, some 8,000 Serbian refugees were resettled in Baranja during 1992 alone.[49] In dilution by pull, a population is ethnically diluted by being pulled out of a region and dispersed elsewhere. In one sweep, a region may be cleansed of undesirables, while at the same time, those undesirables are prevented from concentrating elsewhere. With these goals, ethnic Hungarians were dispersed across Romania, Vlahs and Pomak Turks were dispersed across Bulgaria and the concentration of Greeks in southern Albania was weakened as they were dispersed throughout the country.

An extreme extension of the concept of population dilution is what has come to be known as ethnic cleansing. Bell-Fialkoff has defined ethnic cleansing as "the expulsion of an 'undesirable' population from a given territory due to religious or ethnic discrimination, political, strategic or ideological considerations, or a combination of these."[50] It has recently come to be associated with the war in Yugoslavia, and even more crudely, with the actions of the Bosnian Serb nationalists. Ethnic cleansing has been in operation across the globe since time immemorable.[51] As Bell-Fialkoff points out in his study of the practice, it is "historically speaking neither new nor remarkable."[52] Indeed, in the Americas, the arrival of whites led to attempts at extermination of the indigenous American Indian populations, and when that failed, their containment on reservations. Jews have been expelled throughout Europe and throughout history, culminating with the experience of the mid-twentieth century. Turks cleansed regions of the Armenians, Eastern Europe cleansed their states of Germans, removing over ten million people in one sweep, Stalin resettled entire nationalities, such as the Cossacks, and the Sudanese government has engaged in forced conversion or expulsion of Christians in the southern regions of Equatoria, Upper Nile and Bahr el-Ghazal. In the Balkans, within the memory of living individuals, the Serbs were cleansed from regions of Greater Croatia, at which time some 750,000 were killed (the Serbian revenge resulted in 100,000 estimated deaths).[53] In addition, Bulgarians and Hungarians expelled 120,000 and 70,000 Serbs respectively from the Yugoslav region that they came to occupy. That individuals are alive today that experienced these World War II events may lend some historical dimension and understanding to the atrocities of the Yugoslav war of succession.

As it is not appropriate to identify ethnic cleansing as something that was invented in the present Yugoslav war, so too is it not appropriate to attribute the practice solely to the Bosnian Serbs. While the Serb irregular and paramilitary forces have undoubtedly engaged in much terror against non-Serbs in the recent civil war, they are not alone in their efforts. While Serbs have expelled Croats from eastern Slavonia and Muslims from Eastern Bosnia, so too have Serbs been victims of similar tactics and expelled from Herzegovina and western Slavonia. If Serbs alone were the perpetrators of ethnic cleansing, it is difficult to explain the 600,000 refugees in Serbia, half of which are from Bosnia-Herzegovina. It is also difficult to explain that there are presently virtually no Serbian inhabitants in western Slavonia, or western Herzegovina. The evidence of Croatian ethnic cleansing was clear long before the May attacks against Muslims in Mostar: indeed, Lord Owen suggested in November 1992 that Croatia too should be given sanctions for its role in ethnic cleansing.[54] Moreover, human rights activists in Croatia have been protesting that their government is doing nothing to stop the "well-organized provocations aimed at driving ethnic Serbs out of Croatia".[55] The Yugoslav war is a dirty civil war, in which neighbor has turned against neighbor, and each group is trying be the first to cleanse undesirables lest they cleanse him instead. As *The Economist* aptly put it, it is a war in which the options are to "cleanse or be cleansed."[56]

4. Governments may alter their population composition by adopting a policy of receptive, selective immigration. This open-arms policy serves the purpose of diluting the domestic population by increasing the relative size of one ethnic group. As such, it does not discriminate by skill or familial ties but rather by race, ethnicity or religion. Immigration policies across the world are increasingly exhibiting this tendency: witness the current changes in the Italian immigration law pertaining to immigrants from Africa or the German attempts to relocate Gypsies from their territory. In the Balkans, in-migration is encouraged only when the nationality of the immigrant reflects that of the titular majority: indeed, Greece is not receptive to the inflow of illegal Albanian immigrants; Yugoslavia has strengthened its borders to control the inflow from Albania; Romania and Bulgaria are on the lookout for Gypsy migrants; and so on. Balkan governments do, however, welcome some ethnic groups, as they are perceived as both strengthening the majority (as would result from the Bulgaria's decision to accept some 700,000 Bulgarians from the former Soviet Union) while increasing economic growth (as resulted from the influx of Croat expatriates after the cold war).
5. Governments might alter the composition of their population by engaging in the conquest of territory populated with the desired peoples. Whether this annexation is achieved through war or negotiation, its basis is irredentist sentiment according to which populations divided by international boundaries strive for union. Irredentism is rampant across the globe: in Kashmir, Sri Lanka, North Ossetia, and the Trans-Dniester Republic, for example. Irredentism is (and has been) also present across the Balkans. The clearest examples are among the Serbs in Croatia and Bosnia who want to join Serbia in what has come to be called Greater Serbia, the Croats in Bosnia who want to join Croatia in a Greater Croatia and the Albanians in Kosovo who want to join a Greater Albania. If the Romanians in Moldova and in the Ukraine succeed in their efforts to join their territory to Romania, the size of the Romanian titular majority would increase, as would their power relative to other groups. There is evidence of irredentist sentiment among the Hungarians in Transylvania (and Slovakia and northern Vojvodina) who find themselves on the wrong side of a border.

> In all these cases, a redrawing of boundaries would increase the population of one group relative to others. If to this category are added all those populations that find themselves spread over several countries and desire to unite their territories and create a new country (for example, Spanish and French Basque provinces, or Kurdish regions in Iran, Iraq and Turkey), then the number of irredentist populations increases, as do the possibilities of future demographic power struggles.

The shifting of populations in order to achieve political aims is not a new phenomenon, but it is one that has gained in importance in this new world disorder. Especially in the present atmosphere of self-determination based on ethnicity, the ethnoreligious composition of a population is of utmost importance. However, the importance of numbers in determining both political and economic power of a group relative to others is only valid in a legal system in which rights are based on groups rather than individuals. Such a system is different from one based on the Renner/Bauer model, according to which "national rights should be accorded to individual persons rather than exclusively to territorial groupings."[57] Renner and Bauer wrote about the situation in which "national or ethnic groups were so interspersed geographically that any neat division between 'their' territories was impossible." As long as the ethnic group rather than the individual is the relevant unit within society, there will be pressures to increase relative numbers of people considered desirable. As long as nationalists think that they can achieve self-determination on the basis of the ethnic population of a territory, then they will strive to create an ethnically pure population in the region or regions they covet. The quickest way to achieve these goals is by forcing populations to relocate. Refugees carry with them emotional baggage that can translate into political support for nationalist policies to a degree not possible among peoples not directly touched by this experience. Hence the popularity of the practice of ethnic cleansing. In addition to being despicable and immoral, however, ethnic cleansing carries with it costs that will play themselves out over the long run in both the economic and political spheres.

7

Toward Larger Ethnic and Economic Units in the Balkans

"The need [presently] is to induce de-Balkanization."

Flora Lewis, writing about western Balkan policy[1]

"The international community may actually find it easier to work towards the restoration of some Yugoslav entity, to be brought in some closer association to the EC, than to keep the divided Yugoslavs in their mini-states at peace with each other."

John Zametica[2]

Notwithstanding the dismal mood pervading world opinion in 1993, the war in Yugoslavia will end one day. While an eventual conclusion of organized large-scale hostilities is certain, uncertainty surrounds the form and shape that the future states of the region will take: their boundaries may surprise us, as may their social, political and economic fabric. The future of these states, as well as their relationships, have not been a topic of discussion in the media or academic circles, as attention is focused instead on the process by which the war will end. Indeed, the entire Balkans have been put "on hold" since late 1992, as the world weighs the possibilities of the Yugoslav war continuing, spreading or intensifying. Investment and trade decisions are postponed, long-range planning is discouraged, foreign capital and tourists shy away and risk adversity prevails. Yet a focus on the short term leaves a vacuum at the end of the tunnel, possibly with menacing consequences.

The most compelling reason to think about the long term future of the Balkans is that some form of cooperation is inescapable for the peoples of southeastern Europe. Each of these peoples, with their historic ties to land, have no choice but to coexist and to interact in social, economic and political ways. Indeed, when the Yugoslav war comes to an end, the various nationalities will of necessity resume their relations, albeit in altered form, as they have done after any of the numerous wars in their turbulent, interwoven history: five wars have been fought in the former Yugoslavia this century. Three of those involved Serbs and Croats, on opposing sides.[3] And yet after each war, they overcame their mutual hatred and intermarried, cooperated and traded. Unlike in cases of divorce, secessionist Slovenes, Kosovar Albanians, Transylvanian Hungarians, or Macedonians cannot simply move (and take their land with them, as many might desire—preferably to Western Europe). Abraham Lincoln aptly commented on such situations, "A husband and wife may be divorced and go out of the presence and beyond the reach of each other; but the different parts of our country cannot do this. They cannot but remain face to face."[4] More recently, Cedric Thornberry, commander of the United Nations in Croatia, said that Serbs and Croats ultimately have no choice but to cohabit in a multiethnic and multireligious territory.[5] Even Bosnian president Izetbegovic, taking a dramatic turn from his views of Islamic supremacy as expressed in the Islamic Declaration, recently said that the Yugoslav peoples will live together again, with the help of time and an improved economy.[6] Moreover, the leader of the Bosnian Serbs, Dr. Radovan Karadzic, said that it was time to move on and rebuild ties among the Muslims and Serbs,[7] while Dr. Ljubo Sirc, leader of the Liberal Democratic Party in Croatia, called for improved relations among all the peoples of former Yugoslavia.[8] In light of the post-World War II decades of peaceful coexistence, the resurfacing of centuries-old hatreds said to be at the root of the present Yugoslav war is simply evidence that regional, ethnic and religious intolerance move in cycles with tolerance. Short of a major break with the past, this tolerance will reemerge in the Balkans and with it some form of coexistence will develop.[9]

This chapter contains a study of the future of the former Yugoslav republics and their Balkan neighbors in the aftermath of the Yugoslav civil war. It describes two extreme scenarios, one characterized by the proliferation of numerous small nation-states, and the other characterized by larger multiethnic conglomerates. It explores these two scenarios on both economic and ethnic grounds, evaluating their costs and benefits. It concludes that a large-scale umbrella based on cooperation and tolerance is the only feasible long-term outcome for the region. This chapter does not hypothesize as to the territorial boundaries of the Balkan states in the future: that job is left to the Walter Lippmanns of the 1990s.[10] It furthermore does not propose concrete characteristics of the future cooperation, as embodied in the terms *federation* or *confederation*, but only explores past attempts and possible future directions. To do more at this time would be simple guesswork, given the ambiguity that still envelopes the duration, the complete identity of participants and the final outcome of the Yugoslav war.

It has been shown in the above chapters how previously unlike states, while sharing some characteristics that warranted classifying them as "Balkans," have in the past few years become more similar insofar as they share two pervasive characteristics, economic decline and the rise of nationalism. From their present quagmire, a tendency to perceive fragmentation by ethnic group, with its concomitant territory, as a solution to economic stagnation, political repression and social malaise has emerged. This is the final stage of nationalist bankruptcy, when all other options for ethnic cooperation are exhausted and ethnic purity is seen as a panacea. This is clearly indicated by events in the former Yugoslavia, where Slovenia and Croatia declared independence in June 1991 and were followed soon thereafter by Bosnia-Herzegovina and Macedonia. In April 1992, Serbia and Montenegro declared a new federation under the name of the Federal Republic of Yugoslavia.[11] The Serbian population of Croatia declared an independent Serbian Republic of Krajina (December 1990) and the Albanians of Kosovo declared independence (September 1991), while the Serbs of Bosnia-Herzegovina proclaimed the Serbian Republic (March 1992) and the Croats proclaimed the Croatian Community of Herceg-Bosna (July 1992). Moreover, in September 1993, Fikdar Abdic declared secession from the Bosnian Muslim territories, and proclaimed the "Autonomous Province of Western Bosnia" in the Bihac pocket of Bosnia. This was followed in October by rumors that the Tuzla pocket would also declare its autonomy.[12] Thus one state gave rise to many, of which only three enjoy complete international recognition.

PROLIFERATION OF SMALL NATION-STATES

What would happen if the fragmentation that began in Yugoslavia were to ripple through the entire Balkans and pressures for secession mounted in places like Southern Albania, Transylvania, and Greek Macedonia? In addition, what if pressure for other forms of ethnoterritorial adjustments increased, such as the incorporation of parts of Vojvodina into Hungary or parts of Moldova into Romania? Moreover, what if the international community rushed to recognize the legitimacy of those demands? The result would be a proliferation of numerous ethnic nation-states, of varying sizes and resources, and thus of varying potential and viability: indeed, if all secessionist and irredentist efforts enumerated above were successful, over 20 Balkan states would dot the map. While some writers have identified positive aspects to the creation of microstates, such as the increased ability to sustain peace (notably Kohr)[13] or the increased preponderance for cultural revival (notably Simonton),[14] there would be numerous problems in their proliferation in the Balkans under the present conditions. Clearly, microstates exist and have thrived throughout the world—for example, as Andorra, San Marino, Butan, and Liechtenstein. However, they are usually single microstates existing within the context of larger states with which they have a patron relationship, rather than within a sea of similar microstates.

There are numerous problems of small states, as a look at medieval history may show: the eastern lands under the politically centralized Byzantine Empire achieved a much higher level of economic and cultural development than the small political units that dotted Western Europe in the dark ages. However, today the problems associated with a proliferation of small nation-states are both ethnic and economic. Four concerns that limit the feasibility of such a solution come to mind. The first is the impossibility of drawing clear territorial boundaries in much of the Balkans. The ethnic population of the Balkans is very diverse and in many areas largely intermixed, as indicated chapter 2. In former Yugoslavia, there is significant intermarriage and intermixed residence, especially in Bosnia-Herzegovina. In addition, President Tito created a nationality called Yugoslavs, which according to some post-World War II censuses represented a reasonable choice for some 10 percent of the population.[15] While former Yugoslavia is by far the most heterogeneous of all Balkan states, the situation in the others is also complex. This is due to several factors, such as the refusal of the Greek and Bulgarian governments to acknowledge the existence of some minorities (notably the Macedonians) and the inconsistencies caused by distinguishing people alternatively by religion and ethnicity (notably the Pomaks and Turks, or the peoples of Bosnia). Moreover, there are numerous ethnic groups in the diaspora throughout the region, the most numerous of which are the Serbs outside of the new Yugoslavia and the Hungarians in Romania and Vojvodina.

The second concern focuses on the population movements that would be necessary in order to accommodate the drives for ethnically pure nation-states. While large-scale population migrations are by no means new to history, when involuntary they are accompanied by enormous material and emotional costs. Indeed, the resettlement of Hindus and Muslims after the partition of colonial India, that of Greeks and Turks after the war of 1922 and that of Germans from their traditional lands throughout Eastern Europe after World War II point to the costs associated with large-scale relocation efforts. While the history of the Balkans includes numerous examples of migration resulting from push factors aimed directly at an ethnic group, such an abrogation of human rights in the late twentieth century offends western sensibilities and thus forces controls and compensations on perpetrators that were not part of relocation calculations in the past. Today, the costs resulting from international pressures are immeasurably higher than ever before, as attested to by the Bulgarian government's effort to expel unassimilated Turks in the late 1980s.

In the absence of ethnic purity, a division of the Balkan states into small regions would leave regions with same problems of multiethnic composition, as well as the same economic and political problems. Indeed, according to Allcock, a breakup would only create new states "to confront the same problems of multiethnic composition, an economically marginal situation with respect to Western Europe and the urgent need to modernize and relegitimate their state structures," as was the case in former Yugoslavia.[16]

The third concern relates to democratic rights of minorities in the new nation-states. As indicated in chapter 2, the pursuit of self-determination in Eastern Europe in the late twentieth century has proved not to be a universally democratic choice. While there is a way in which self-determination seems most democratic, since it allows the individual to determine which country he or she wants to be a part of (even if, in the limit, the population shrinks to one), in an atmosphere of rabid nationalism, it creates a situation in which democratic rights are conferred by ethnic group, at the cost of nonmembers. According to Etzioni, "The great intolerance breakaway states tend to display toward minority ethnic groups heightens polarization [among groups]."[17] Indeed, Hayden points out that the new successor states of Yugoslavia all have constitutions that are less democratic with respect to minority rights than the last Yugoslav federal constitution before the breakup.[18] These undemocratic tendencies have emerged when minorities become majorities and majorities become minorities, since, as Jaszi points out, "the political morals of an oppressed nation change when it comes to power."[19] This is supported by Harris, who claims there has been a proliferation of "perverse national liberations," as there are cases of "hideous oppression of one minority is translated into the hideous oppression of another," such as in the case of the Boers (who were mistreated by British and then went on to mistreat South Africans).[20] As Pfaff pointed out, "the idea of the ethnic nation thus is a permanent provocation to war."[21]

A fourth concern pertains to the economic potential of small nation-states. Fragmentation implies the creation of regional units that must deal with the loss of benefits associated with size, such as economies of scale and a varied resource base. The resulting loss of markets increases dependency on the outside world and costs due to duplication of services and inefficiency.[22] In addition, the new nation-states incur costs embodied in the creation process, as they face the daunting prospect of setting up independent institutions (such as currencies, central banks, legal and tax systems, et cetera) to enable independent functioning of their economies. Moreover, if fragmentation is occurring in a previously communist state that is undergoing a simultaneous transition to a market economy, then this structural change also imposes its short-term costs on the new economy. These sources of economic change are simultaneously convulsing the successor states of Yugoslavia, obscuring a clear assessment of viability of the newly emerging states. Under the best of circumstances, economic obstacles embodied in any one of these three changes may be hard to surmount. Under circumstances of war and sanctions, viability is all the more questioned. Of the newly created Balkan states, Slovenia seemed to have the best prospects for economic prosperity as an independent entity as a result of the high level of economic development prior to secession. However, it too has suffered from the loss of markets in Yugoslavia and the disappointing flow of foreign investment, as have Macedonia and Croatia.

LARGE, HETEROGENEOUS UNITS

Instead of the creation of small ethnic enclaves, let us explore what would happen if that trend were reversed and states were encouraged to expand to include more territories and ethnic groups. Such an integrative trend is evident among both formerly divided regions (such as Germany), formerly united regions (such as the Commonwealth of Independent States) and among formerly unrelated states (such as the EC, the Visegrad Four and the Central European Initiative). Such an integrative trend has been called for by Zametica, who has written that a unified Yugoslavia does not seem so bad after all, and by Lewis in her discussions of the Yugoslav war that "all proposals to separate combatants . . . would only advance Balkanization, the region's historic handicap. The need is to induce de-Balkanization."[23]

Large heterogeneous states would have the advantage of incorporating numerous ethnic groups of the region into a political superstructure, thus institutionalizing heterogeneity and deriving strength from size and diversity. Contrary to the perception of Roucek, who wrote of "the handicap of heterogeneity," ethnic diversity may be viewed as an asset.[24] The internal interethnic relations of larger states could be based on a variety of models, with consideration for the size of ethnic groups coupled with a clear expression of ethnic rights. Such an arrangement has some advantages. First, it would lend protection to the smaller ethnic groups currently disadvantaged by nationalist constitutions. It is a fact that all Balkan ethnic groups, as a result of their histories of conquests and repression, live in fear of the moment their ethnic groups become a minority, because as Ra'anan pointed out, "when minorities become majorities, their political morals change."[25] Second, it would control the mutually exclusive aspirations for the restoration of former territorial glory, such as Greater Bulgaria, Greater Serbia, Greater Albania, et cetera. Third, it would give a homeland to those peoples that presently find themselves without territory, namely the Romas and Yugoslavs.[26]

The economic advantages of larger heterogeneous units are many, including economies of scale, extended markets and consequently enhanced viability. These benefits are best exploited when states have economic links that transcend mere trade. Indeed, trade between many small states represents an expensive duplication of services and institutions. It also implies likely protection policies that hurt consumers and ultimately producers. There is no reason to assume that cooperation among these trading partners would take place without some other level of cooperation, such as common economic policies.

In the Balkan states, with their current ethnic polarizations and pressures for fragmentation, there is a belief that new nation-states will survive and thrive in the absence of relations with their neighbors. Yet the Balkan states, whatever their boundary configurations throughout history, have always engaged in economic interactions, ranging from trade to joint ventures to tourism.[27] These relations have actually intensified since 1990 and will continue to do so, as a result of several factors. First, post-World War II trade patterns of Balkan members have changed. The demise of Comecon has forced Romania and Bulgaria toward their neighbors. Indeed, Romania has been especially harshly affected by the disruption of Comecon links, such that its trade with Serbia increased to about $400 million in 1991.[28] The Albanian economy has opened up, as trade is seen as the way to prevent economic disintegration. However, while Italy is its preferred trading partner, its lukewarm response to Albanian overtures is forcing Tirana to turn eastward. The increased marginalization of Greece within the EC, intensified by the 1993 creation of a single market and the demise of a bipolar world, will be revealed when the mask of sanction-related trade disruption is lifted from trade accounts. Second, western preoccupation with internal issues and domestic economies, coupled with an overabundance of potential investments in the former Soviet Bloc, has led to less than enthusiastic interest in the Balkan states. This sentiment is intensified by dire predictions that the Balkan economies are less likely to succeed than their northern neighbors such as Poland, Hungary, the Czech Republic and even Slovakia. The effect of these predictions is exemplified by the pattern of German investment in Eastern Europe, which, as pointed out by PlanEcon, has reached its pre-World War II level and is significantly more pronounced in the northern states than in the Balkans.[29]

All these reasons constitute incentives for Balkan economies to turn toward each other. Yet during the past three years, Balkan economic interactions did not reach their full potential because of the Yugoslav wars. The effect of these wars on various forms of economic interaction has been pronounced, including in the areas of transportation and shipping routes (the loss of which is felt clearly in Macedonia), tourism (the loss of which is felt clearly in Croatia), flows of resources into less-developed regions (the loss of which is felt in Kosovo) and trade along the Danube (the loss of which is felt in Hungary, Romania, Ukraine and Bulgaria). The effect on Balkan trade of the imposition of sanctions on Yugoslavia is ambiguous insofar as it both disrupted legal trade while at the same time forced Serbia and Montenegro to turn to their immediate neighbors for petty sanction busting.

Thus, given the importance of a variety of neighborly economic links to the Balkan states, it would be wise to plan on solidifying these links in the aftermath of war and sanctions. Economic relations among numerous small nation-states such as those that might arise in the Balkans would rapidly spiral into ethnically based competition and regional economic demands aimed at improving one ethnic group's position relative to another. However, economic relations that take place under a multiethnic umbrella might avoid such pitfalls while enjoying the economic benefits of size.

BALKAN COOPERATION

Balkan Federation in Historical Perspective

Serious attempts to achieve cooperation and integration of the Balkan states have been proposed over a long historical period. Giannaris points out that during Ottoman rule, such attempts were made by Rhigas, Ypsilanti and the

Pulios brothers; during the 1920s, by Papanastassiou, Milonas and Stephanopoulos in Greece, Radic in Croatia, Stambuliski in Bulgaria and Titulescu in Romania; during the 1960s, by Mercouris in Greece and Gonofski in Bulgaria; and by numerous other leaders throughout history, including perhaps most strenuously the Yugoslav communists after World War II.[30] But what was in fact achieved in these attempts at cooperation and integration, and how was reality different from the revolutionary ideals of those such as Garibaldi and Mazzini? In ancient times, Philip and Aleksander of Macedon did succeed in establishing a Balkan Hellenic state, but it disintegrated after their deaths. The first concrete example of a Balkan union in modern times is the Balkan League, consisting of Serbia, Romania, Montenegro, Greece and a Bulgarian revolutionary society. It was organized by the Serbian prince Michael III, with the goal of expelling the Turks from the Balkans and uniting the south Slavs in a single state. Serbia was to be the Piedmont of the Balkans. However, the existence of the league was tied to Prince Michael, and upon his assassination, the integrative efforts died, giving the league a mere two year existence (1866-68). A new Balkan League was established in 1912; this time the members included Serbia, Bulgaria, Greece and Montenegro. The goal was not only to expel Turks from the Balkans (specifically from Macedonia), but also to combat Austria's rising influence in the region. The League disintegrated when the allies won the First Balkan War and then disagreed on the division of the spoils. This led to the realignment of the allies and the Second Balkan War in 1913, which was fought against the Bulgarians and led to the division of Macedonia between Serbia and Greece. During the interwar period, integration was achieved on a small scale, as the former Austro-Hungarian dominions of Slovenia, Croatia, Slavonia and Dalmatia joined Serbia and Montenegro to form the Kingdom of the Serbs, Croats and Slovenes, thus uniting some segments of the southern Slavs. Renewed efforts at Balkan integration occurred during World War II, based on the principle of "the Balkans for the Balkan Peoples." Stavrianos, in his thorough study of Balkan federation efforts, notes the machinery that was developed for the Balkan Union, which was proposed to go into effect after the war and was given formal standing by the Greek-Yugoslav agreement of 1942.[31] The machinery was to include a political, economic and military organ, to be subservient to a Permanent Bureau. In the aftermath of the war, the Yugoslav communists continued to insist on the idea of federation, proposing even to include Albania and to resolve the question of Macedonia. However, plans for this Balkan Union also disintegrated as new alliances grew out of World War II and were strengthened when communists failed to come to power in Greece, which joined the European Community and NATO, while other former Balkan League states were pulled into the Soviet orbit.

In all of these movements towards integration, there was always concern about large foreign powers. Indeed, the basis for unity was always the threat or actuality of foreign intervention, be it by Austria, Turkey or Germany. Nor did Russia refrain from meddling in the Balkans. At the time of the first Balkan League, the overriding sentiment was anti-Turkish, and during the Balkan League of 1912, the sentiment was anti-Austrian and anti-Turkish, with the diplomatic blessing of Russia. Thus foreign meddling in the Balkans produced efforts to unite the Balkan peoples in order to protect them from the outside. Now, in the 1990s, the threat of foreign powers is present, albeit less obvious. While the biggest threat to the Balkans is from within, there is no doubt that pressure from outside powers has exacerbated a delicate situation. The roles played by Germany and Austria have been strongest, reviving memories of two past efforts at encroachment into the region in this century alone. Indeed, the Polish monthly *Waterplatte* has recently said that German policy toward Eastern Europe in general is a new version of the former "Drag Nach Osten,"[32] while a poll taken across Western Europe indicates that on average, over 30 percent of the population perceives unified Germany to be a threat to peace (in Greece, that number is close to 50 percent).[33] Germany's diplomatic activity on behalf of Croats, culminating in its recognition of Croatia, with which it broke ranks with its EC allies, is just part of its meddling policy in the Balkans. These efforts have also included the negative publicity that demonized Serbs, leading Conor Cruise O'Brien to state, "The Croats get better press than the Serbs, not because they deserve it, but because they have a powerful patron inside the [European] Community and the Serbs do not."[34] Moreover, there is clear evidence of German busting of the weapons ban in the Balkans and supplying of arms to the Croatian government.[35] Neither has Turkey been silent in the dispute. Ostensibly in its efforts to come to the aid of the Bosnian Muslims, Turkey has taken an anti-Serbian approach whose historical roots are found in the role Serbia played in the demise of Ottoman influence in the Balkans. Turkey has been active in the present efforts at organizing a Balkan response to the Yugoslav war, embodied in the Balkan Conference in Istanbul; however, its focal point has been the isolation of Serbia.[36] Turkey's view of the Balkans was clearly spelled out by President Turgut Ozal in his 1990 publication calling for a revived Turkish role in the Balkans.[37] The intervention of the international Muslim community, while with a goal similar to the Turks', is different insofar as the Muslim countries do not have traditional ties to any of the combatants in the civil war but rather religious ties to one of the players. Nevertheless, they have contributed by causing pressure within international bodies and on foreign governments, by contributing money, weapons and volunteer fighters to the Bosnian Muslims, and by offering to send arms in the event of the lifting of the ban on weapons. Also of importance is the involvement of Russia in the Yugoslav conflict, albeit this involvement is not to the extent that the Serbs would like. This traditional ally of Serbia has been Serbia's potential supporter more than its actual backer: indeed, its consistent voting at the United Nations in resolutions against Serbian interests have been more pronounced than its efforts to lift the sanctions. In Western Europe, the sense is that traditional ties of the great powers in the Balkans cannot be forgotten; however, while Germany so strenuously supports its traditional ally, the Croats, the prevailing popular anti-Serbian sentiment prevents France and England from supporting their traditional ally, namely Serbia. The United States has been drawn into the conflict despite itself, in part as a response to public outrage. Without a clearly defined policy, both the Bush and Clinton administrations have coped with public pressure by making ad hoc moves against the Serbs, succumbing to forming foreign policy on the basis of media, which, for its own motivations, has portrayed a highly complex situation in terms of black and white, seeing the Serbs as aggressors and the Croat/Muslims as victims. The U.S. actions—the imposition of sanctions on Serbia and later their enforcement, as well as the lead they took in diplomatic efforts to isolate the Serbs—all point to an active intervention in the war. To a lesser degree, Hungary took sides at the onset of the war, reviving its historical ties to Slovenia and Croatia. It has put its bias into action by strongly enforcing the sanctions against the new Yugoslavia and by selling arms to the Croats despite the arms ban. Moreover, it has incited Hungarians in Vojvodina (as well as Slovakia and Transylvania), masking the fact that despite the passing of almost a century, they have not forgotten that they lost territory to the Serbs (and others) after World War I. Such strong support by the Hungarian government in Budapest resulted in a plea from the Vojvodina Hungarian Party to decrease its involvement because it was becoming counterproductive. Albania too has historical reasons for taking sides: it recalls that Serbia opposed its creation, its boundaries and its very existence. Today, Albania is taking a strong stand in siding with not only the Kosovar Albanians, but also the Bosnian Muslims in Bosnia and Sandzak.

Thus arguments that the present situation in the Balkans is not internationalized and is free of big-power intervention are simply false. Big powers, while not fighting with their armies, are exerting their influence in other ways. I beg to differ with Lewis when she argues that outside powers are currently not enemies and have a common interest in the Balkans, namely peace and security. The inability of the western powers to agree on a Balkan policy came to a head in May 1993, indicating that outside interests in the region exist—that while the outsiders are not enemies as they were in the past century or at times in this century, nevertheless they are not able to achieve unity among them.

The Nature of Balkan Cooperation in the Twenty-First Century

In the short run, the overriding goal of the newly formed states of the Balkans is sovereignty. Once it is achieved, as it has been in Slovenia, Croatia and Macedonia, other goals predominate in importance, such as those encompassing issues of economics, security and crime. Indeed, Slovenia joins Lithuania as a country whose independent government and political coalition has not survived in the post-independence period. However, in the long run, the focus changes and often turns to a new desire for normalization of relations with former union states. This occurred in many parts of the world: President Rahman of Singapore at the time of the break between Singapore and the Malaysian Federation claimed that there may be prospects for reunification in the future, after a taste of sovereignty, since "absence makes the heart grow fonder."[38] This view was also voiced by the Ukrainian minister of defense: "What we need today is a chance to breathe some freedom, to really feel some sovereignty. Then maybe in five years or so we will start talking of uniting or some association."[39] Already, with the Yugoslav war not even over, Croats have expressed a desire for renewed economic ties with former Yugoslavia: in May 1993, a poll was taken in Croatia to determine the popular views on future relations with Serbs, and almost 60 percent favor some kind of "loose customs and economic links between Croatia and rump Yugoslavia."[40]

It is the exact nature of this association that is to be explored here. If the goal is to reverse the damage of nationalist bankruptcy by creating a political system that would enable the Balkan populations to live together in stability and peace, and an economic system that would enable the Balkan populations to achieve the highest rate of economic growth and the most rapid recovery from the effects of the war, sanctions, refugees, the transition to capitalism and the end of the cold war, then an appropriate form of cooperation might be a confederation. A confederation is a loose joining of sovereign states that includes elements of economic, political and military association. It is defined as "a voluntary association of independent states that, to secure some common purpose, agree to certain limitations on their freedom of action and establish some joint machinery of consultation or deliberation."[41] While this definition leaves wide open the extent of the limitations and cooperation, in the particular case of the Balkans, it might entail the following: With respect to the economic aspects, it would ensure that the sovereign Balkan states have at least a customs union and perhaps even a common market.[42] In other words, they should abolish internal tariffs and have a common trade policy with the outside world. Perhaps as part of a long term goal of common market, member states might add the free flow of factors of production among the member states, although in the short run this is neither practical nor necessary. There should also be a common economic plan for reconstruction, which would focus on building regional relationships and regional economic ties. With respect to the political aspects of a confederation, all interaction should be based on the necessity of compromise, in view of the new world conditions and their negative impact on the Balkan states. With respect to defense, all member states should retain their own defenses, although a common policy toward the outside world is encouraged and a common defense pact against outside intruders may be a viable option to try to break the historical pattern in which the Balkan states have always been merely pawns of greater powers.

The possibility of a confederation seems unlikely in the unruly world of the Balkans, a world in which neighborly relations are often hostile. Indeed, only Greece and Serbia were never at war with each other and were allies in the major wars of this century. Why, in such a world, would countries that are essentially free of minorities, such as Greece, Slovenia and Albania, want to become involved in a multiethnic union whose participants have a history of interethnic animosities? Why would more developed regions, such as Slovenia and Greece, want to become tied to regions that are less developed? Why would Muslim populations, with Muslim support worldwide, want to increase their links with non-Muslims in unions in which they would be minorities? Why would Croatia and Slovenia, with the backing of the strongest power in Europe—Germany—engage in any compromises such as those necessitated by a union with other Balkan states? The answers to these questions are based on the view that those decisions that are made by the Balkan leaders, and not imposed from the outside, are based on rational choice and cost-benefit analysis of the future. Thus any one of the concerns raised above would be counterbalanced by the positive aspects of joining a union. Some concerns are listed below. All of them together are reasons why a mere trading agreement among the Balkan states is insufficient to offset the damage of nationalist bankruptcy in the region.

First, with respect to ethnic issues, all those states that have ethnic populations outside of their borders may be motivated to be part of a union so as to be better able to ensure that the rights of these populations are respected (this includes Albania, Serbia, Greece and Bulgaria). Also, ethnic nationalist revivals and ethnoterritorial aspirations may be dissipated in the context of a larger union (including the Greece-Macedonia and the Kosovo disputes). Moreover, regions may find that being a part of a larger union may increase the possibility of retaining unchanged borders. This is especially clear in the case of Croatia, which might be allowed to keep Slavonia and Krajina if it were part of an umbrella organization.

Second, with respect to economics, a confederative union would convey to the international community of investors that there is a commitment to stability in the region, thus increasing the chances of investment and aid. This may counter the trend explained in chapter 3, namely the West's reluctance to invest in the region, so that the Balkan regions have received less in investment than was expected. A confederative union would also increase the scope for economic interaction among member states, with respect to both large-scale, intraregional investment projects and trade. Even the members of a disintegrating Yugoslavia agreed as a first step and as the minimum to building further economic cooperation in the context of the EC-sponsored Yugoslav peace conference "to establish a customs union with other institutional arrangements leading possibly to an economic union."[43] Some form of reintegration of formerly integrated regions is not unique to former Yugoslavia; indeed, signs of it are already evident in the former Soviet Union, where numerous newly independent states have "volunteered to transfer sovereignty to Russia in the hope of reviving their economies through reintegration with it."[44] A form of reintegration of former Yugoslav lands has been called for and predicted by many individuals: Yugoslav businessmen such as Boris Vukobrat,[45] writers such as the author/former dissident Milovan Djilas,[46] former politicians such as Milan Panic,[47] British government officials such as Mario Viller,[48] historians such as Christopher Cviic[49] and economists such as Branko Horvat.[50] Indeed, the last claims that in the absence of some form of economic integration in the Balkans, the population of his native Croatia is bound to be like that of the Caribbean states—"waiters on the outside, peasants on the inside." Even the current president of Montenegro, Momir Bulatovic, has said that the future must consist of the normalization of relations between former Yugoslav republics.[51] It is in this respect that the more developed regions have the most to gain, since their products would have a ready-made market, while the less developed regions would have access to less expensive goods than those offered by the global markets. For Serbia, a union would have the effect of overcoming international isolation and the quickest recovery from the pain associated with sanctions. Indeed, Serbia faces the daunting task of breaking into new international markets from scratch, which would be facilitated if trade relations among its neighbors were already assured. Similarly, Macedonia would benefit greatly from renewed ties with both Greece and Serbia, and the development of new intra-Balkan transportation routes to both the Mediterranean and Black Seas would encourage prosperity.

Third, a confederate union would temper Islamic fundamentalism in the region. Islamic fundamentalism has come to play an important part in the perceptions of the Balkan populations *as a result* of the war in Bosnia-Herzegovina: the Muslims have become more Muslim, while the non-Muslims have become more distrustful and intolerant of Muslims. This is best exemplified by the case of the Bosnian Muslims, who as a population largely did not adhere to most rules of Islam in the past, but only began to do so when the war with the Christians forced them to renew their identity and to differentiate themselves more. As a result of the loss of territory associated with the civil war, President Izetbegovic has proclaimed that Muslims will wage guerrilla war, as suggested to him by PLO leader Yasser Arafat.[52] The Muslim populations may reject inclusion in a larger union, in part because of the necessary secularity of such a union: while there is no doubt that the Muslim states have become more Muslim as a result of the conflict with non-Muslims, their fundamentalism is not a characteristic that would or could predominate in an atmosphere of peace. Thus a confederative union would decrease the need for extremist Islam in the Balkans and would protect the European Muslims from the pressure to polarize that might come from non-European Muslims. The other side of the Islamic coin has to do with the perceived threat to the Christian populations of Islamization. To this end, some Balkan states might see a union as a way of countering the increasing Islamic threat from both within the Balkans and without. From within, the rising birth rate among the Muslims populations, such as the Albanians in Kosovo, plays an important role in the changing demographics and thus political power in the region. With respect to the external threat, the possibility of a Muslim crescent reaching from the Mediterranean all the way to the Great Wall of China plays itself out in the Balkans, much to the fear of the Orthodox populations. This fear is based not only on greater population growth among Muslims than other ethnic groups of the region, but also on the evidence of aid and investment from the international Muslim community.[53] Greece would be cut off from Europe, while Serbia perceives threats similar to those of the Ottoman Empire. The Catholic regions to the northwest, as well as Romania, have historically been more immune to this Muslim threat.

Fourth, a union would offset the marginalization of the region that resulted from the end of the cold war. Greece is becoming dispensable to the EC as the West no longer needs a foothold in the Balkans; Romania and Bulgaria are no longer in the Soviet orbit, a fact that burdens them with costs as well as benefits. Albania is ready to join the world and drop its attempts at self-sufficiency. The creation of a Balkan confederative union, with political, economic and military ties, would convey a strength of the region and a determination that it will no longer be fragmented by the great powers.

Fifth, a confederative union might be perceived as satisfying some of the same goals as the integration in Western Europe. To make the analogy: if one perceives the Maastrich treaty as a way of controlling a major power in the region, namely unified Germany, then a Balkan confederation would succeed in controlling all those threatening states that others perceive as such. For example, an expansionist Serbia, Croatia or Turkey might be thus controlled.

It is Slovenia that has the strongest arguments against the joining of a Balkan union. It has no ethnic ties to the region, such as a large Slovenian population living outside of its state, and its physical proximity to Western Europe makes it somewhat isolated from events in the Balkans. However, isolation does not mean insulation. Slovenia's economy has been greatly disrupted by the war and other destabilizing events in the Balkans, since it depended on cheap raw materials from the less-developed regions to the south and east for its inputs into production, and it depended on those markets for the sale of its exports. Thus its motivation for union with its eastern members may be purely economic, and if union is the condition for the reestablishment of those trade ties, then it may be worth it. Perhaps an "observer status" might be a short-term solution. Greece, although not clearly more developed than Slovenia, also has an economic incentive, as well as a strong link to the Western markets. With such ties, Greece perceives itself as the agent for the West doing business with the Balkans, especially as the link to the Balkans from the EC.[54]

How can a Balkan union succeed? Are there not too many divergent interests, conflicting extra-Balkan alliances, and too much deep-rooted distrust to even entertain such cooperation?[55] Perhaps. According to Stavrianos, there are two preconditions for the creation of some kind of union in the Balkans: similar political systems, preferably democratic, and the development of a system of collective security. The first condition means that the political systems are similar in all countries, so that common structures and mechanisms exist. The second one means that there is no infringement by outside powers on the member states. As indicated in chapter 3, the condition of a similar system has presently been met in the Balkan states. Indeed, since 1989, the Balkans have embarked upon a path that has increased their similarities more than any other time during their post-World War II history. Presently, for numerous internal and external reasons, the Balkan states have undergone similar pressures and seem to be responding to these in similar ways, mainly with the rise in nationalist leaders operating in a system of semidemocratic institutions. The degree of democracy differs, especially in Greece and Slovenia, the two fringe states in which the democratic process is perhaps most developed. With respect to the condition of collective security, Stavrianos meant the need to ensure that there be no infringements on the security of the region by outside powers, so that the Balkan countries are not forced to seek protection from the outside powers, who in turn must not go to war in the Balkans over the establishment of spheres of influence. The evidence is unclear as to whether this condition has been met. At the time of this writing, there is no doubt that there has been international intervention in the Yugoslav crisis. However, the inability of the world to come to an agreement on the policy to follow to end the Yugoslav war indicates the unwillingness of any great power to go to war over their sphere of influence. There are still strong links between the Balkan states and the outside powers that support them, but these increasingly seem to be waning, and thus it is likely, if this trend continues, that the obstacles to a confederation will be cleared in the future.

A discussion of possible cooperation among the Balkan peoples has to take into consideration the existence of a common culture and identity. Indeed, there has been much debate on the nature of a Balkan identity. The Romanian historian Iorga claims that there is a Balkan identity, one that has been formed by geography, history and climate, and culminates in a "certain unity, a unity which is basic, intimate and profound and which the superficial phenomena of discord, unfriendliness and conflict must not hide from us."[56] Where Yugoslavia is concerned, this is supported by Ignatieff, who claims that the Serbs and the Croats have more that links them than that divides them.[57] It is also supported the Austrian writer Peter Handke, who says that for him, there will always be a Yugoslavia; he claims that in his extensive travels and contacts with all nationalities of Yugoslavia, he never witnessed interethnic animosity: "I have never before noticed the dream about independent Slovenian and Croatian states. It appeared only when Yugoslavia grew very weak."[58] The view expressed by the Croat nationalist leader Kukuljevic in the nineteenth century is shared by many urban and educated people: "Byzantium and Rome succeeded in separating the Serbs and the Croats, but the fraternal tie which unites them is so strong that henceforth nothing in the world will be able to sever it."[59] There are those, of course who counter this view. Nationalists on all sides in the Yugoslav civil war recently claimed that the groups they represent are fundamentally different, in order to create more distance between them and their war enemies. This anti-integrative view is supported by Kaplan, who implied that the real iron curtain in the Balkans should not have been where it was: "The cold war and the false division of Europe were over. A different, more historically grounded division of Europe was about to open up. Instead of democratic Western Europe and a communist Eastern Europe, there would now be Europe and the Balkans."[60] In his view, Slovenia and Croatia belong in Europe.

CONCLUSIONS

After the Yugoslav war, the ethnic groups of the Balkans must turn to each other, even if only out of necessity. On economic grounds, invigoration of their economies will take place through the development of new trade routes among each other, as these constitute, after all, their most logical markets under the competitive circumstances of the 1990s. On the ethnic front, tolerance and coexistence are cheaper and more viable than the creation of ethnically pure regions, whether by war or organized migrations. Therefore, the pursuit of self-determination based on ethnicity should be discouraged and solutions to the economic and nationalist crisis must be found amidst possibilities of larger, heterogeneous units.[61]

Cooperation among the Balkan peoples may be perceived as too optimistic a goal. Such a program might be called infeasible on the basis of the nature of the Balkan population, which is perceived as backward, passionate and with stronger roots in history than reality. George Brock's definition of a nation seems to fit the Balkan populations aptly: "A nation is a people united by a common dislike of its neighbors."[62] However, anything is possible in the Balkans, as evidenced by the very diverse and often ingenious paths the Balkan states took in the post-World War II period. These populations may yet surprise us again. They may yet be able to stand aside and take a long-term and broad view of their economies and demographics, and opt for some dispassionate, rational solutions. They may do this because the enemy, nationalist bankruptcy, is rising from within.

Alternatively, if this fails, then the international community might take on a unifying role. Perhaps Stavrianos was right when he claimed, "Unity in the Balkans is more likely to be forced from without than to arise from within," concluding that "a voluntary Balkan federation is contrary to past experience."[63] Lewis recently argued for a "19th century solution," one that would be imposed from outside on the basis of a grand nineteenth century–style conference, including the United States, the Balkan states, Europe and Russia, that would determine the larger picture in the Balkans.[64] In that case, the United Nations may find itself in yet a new role in the 1990s. Such a role might be supported by President Clinton's use of economic carrots and sticks to achieve goals in the Balkans. The economic incentives to offer Balkan ethnic groups might include large-scale projects, such as development of tourism, communications and transportation—in other words, projects that are regional in nature and extent instead of merely on a country-by-country basis. Economic incentives might be linked with a view of the region that is multiethnic, rather than indulging ethnic groups in their efforts to break up, as the West did in 1991. Indeed, Dusko Doder said that the future of the region should be characterized by economic reconstruction efforts rather than military intervention.[65] Zametica was on the right track when he concluded that "the international community may actually find it easier to work towards the restoration of some Yugoslav entity, to be brought in some closer association to the EC, than to keep the divided Yugoslavs in their mini-states at peace with each other."[66] A solution along these general principles may be what is needed to offset the two major forces presently tearing the Balkans, economic decline and ethnoterritorial nationalism.

NOTES

Notes to Chapter 1

1. *The New York Times,* September 23, 1992.
2. Josef Joffe, "Bosnia: The Return of History," *Commentary,* 94, no. 4:29.
3. Misha Glenny, *The Fall of Yugoslavia* (London: Penguin Books, 1992), 86.
4. Ibid.
5. Gerald B. Helman and Steven R. Ratner, "Saving Failed States," *Foreign Policy* 89 (Winter 1992):3.
6. Anthony Smith, "Chosen Peoples: Why ethnic Groups Survive," *Ethnic and Racial Studies* 15, no. 3 (July 1992):450. Also, note that a nation and an ethnic group are used interchangeably outside of America. In America, a nation is synonymous with a state.
7. John B. Allcock, "Rhetorics of Nationalism in Yugoslav Politics," in John B. Allcock, John J. Horton and Marko Milivojevic (eds.), *Yugoslavia in Transition* (New York: Berg, 1992):287.
8. Its importance is elevated when two or more religions coexist and injustice, whatever its root, is perceived as aimed at people on the basis of their religion. Religion tends to become a focal point in nationalism when religious intolerance is a policy at the governmental level and efforts aimed at religious purity are undertaken. Examples of current nationalist movements that are religious in orientation are: the southern Sudanese rebel groups, which are Christian in a state populated by a Muslim majority, and the Karen, who are Christians, while Myanmar's official religion is Buddhism. The role of religion is even more poignant in the cases of the Catholics and Protestants of Northern Ireland and the Catholics and Eastern Orthodox in the Yugoslav civil war of 1991.
9. Indeed, increased autonomy or independence is demanded when a group perceives the demise of minority culture as exemplified by language. Quebec comes to mind as a region in which extreme effort has been put in reviving French as the regional language, perhaps to the detriment of non-French speaking inhabitants of the region. Catalonia is also revising regional laws to grant precedence to its language in the face of the increasing popularity of Spanish. In both of these cases, language preservation, together with the culture that is embodied in it, is cited as a major reason for the desire for secession.

10. Tom Nairn, *The Break-Up of Britain* (London: New Left Books, 1977).
11. Milica Zarkovic Bookman, *The Political Economy of Discontinuous Development* (New York: Praeger, 1991).
12. William Beer, *The Unexpected Rebellion: Ethnic Activism in Contemporary France* (New York: New York University Press, 1980).
13. Michael Hechter, *Internal Colonialism: The Celtic Fringe in British National Development 1536-1966* (Berkeley: University of California Press, 1975).
14. Peter Alexis Gourevitch, "The Emergence of Peripheral Nationalisms: Some Comparative Speculations on the Spatial Distribution of Political Leadership and Economic Growth," *The Comparative Study of Society and History* 21 (July 1979): 303-322.
15. James Simmie and Joze Dekleva, *Yugoslavia in Turmoil: After Self-Management?* (London: Pinter Publishers, 1991): xvii.
16. Egon Zizmond, "The Collapse of the Yugoslav Economy" *Soviet Studies* 44, no. 1 (Winter 1992): 110.
17. The relationship between nationalism and economics is often so strong that the two can be difficult to separate in judging the source of ethnic activity. Indeed, how can we distinguish between the desire for national control of resources among the Azerbaijanis or the inhabitants of diamond-rich Yakutia and the pride in their culture and the desire to see their people in power?
18. See Anthony Smith, *The Ethnic Revival* (London: Cambridge University Press, 1981), for a description of this view (pp. 1-3).
19. Karl Deutsch, *Nationalism and Social Communication,* 2d ed. (Cambridge: MIT Press, 1966), and *Nationalism and Its Alternatives* New York: Knopf, 1969); Samuel Huntington, *Political Order in Changing Societies* (New Haven: Yale University Press, 1968).
20. This view is attributed to Otto Bauer, and discussed in Alexander J. Motyl, "From Imperial Decay to Imperial Collapse: The Fall of the Soviet Empire in Comparative Perspective," in Richard L. Rudolph and David F. Good, eds., *Nationalism and Empire: The Habsburg Empire and the Soviet Union* (New York: St. Martin's Press, 1992), 28.
21. This view is held by Michael Hechter in *Internal Colonialism* and Tom Nairn in *The Break-Up of Britain.*
22. Walter Connor, "Nation Building or Nation Destroying?" *World Politics* 24 (1972): 344.
23. Miroslav Hroch, *Social Preconditions of National Revival in Europe* (Cambridge: Cambridge University Press, 1985); Beth Michneck, "Regional Autonomy, Territoriality, and the Economy," paper presented to the American Association for the Advancement of Slavic Studies, Washington, October 1990; Anthony Birch, *Nationalism and National Integration* (London: Unwin Hyman, 1989).

24. Christine Drake, *National Integration in Indonesia: Patterns and Policies* (Honolulu: University of Hawaii Press, 1989), 145.
25. Immanuel Wallerstein, *Africa: The Politics of Independence* (New York: Vintage, 1961), 88.
26. Milica Zarkovic Bookman, *The Political Economy of Discontinuous Development* and *The Economics of Secession* (New York: St. Martin's Press, 1993).
27. Aleksander Gershenkron, *Economic Backwardness in Historical Perspective* (Cambridge: Harvard University Press), 29.
28. Ernest Geller, "The Dramatis Personae of History," *East European Politics and Societies* 4, no. 1 (1990): 132.
29. Victor Zaslavsky, "Nationalism and Democratic Transition in Postcommunist Societies," *Daedalus* 121, no. 2 (Spring 1992): 117.
30. Eric Hobsbawm, *Nations and Nationalism Since 1780: Programme, Myth, Reality* (Cambridge: Cambridge University Press, 1990) 164, 177.
31. J. L. Talmon, *The Myth of the Nation and the Vision of Revolution* (London: Secker and Warburg, 1981), 544-45.
32. This phrase was first introduced by *The Economist*, February 6, 1993, p. 53.
33. Peter Leslie, "Ethnonationalism in a Federal State: The Case of Canada," in Joseph Rudolph and Robert Thompson, eds., *Ethnoterritorial Politics, Policy and the Western World* (Boulder: Lynne Rienner Publishers, 1989) 47.
34. Ian Bremmer, "Fraternal Illusions: Nations and Politics in the USSR," paper presented to the American Association for the Advancement of Slavic Studies, Miami, Florida, 1991, 47; Albert O. Hirshman, *Exit, Voice and Loyalty* (Cambridge: Harvard University Press, 1970).
35. Sometimes there is also the element of revenge, such as in the expulsion of ethnic Germans from Czechoslovakia or Italians from Yugoslavia after World War II.
36. Connor, 320.
37. Charles Gati, "From Sarajevo to Sarajevo," *Foreign Affairs* 71, no. 4 (Fall 1992).
38. Given the ethnic link between the Azerbaijanis and the Turks, the central government in Istanbul announced in March 1992 that it would be difficult for Turkey to not intervene on behalf of their fellow Muslims. *The New York Times,* March 7, 1992.
39. India's involvement began when it retaliated for the hijacking of an Indian Airlines airplane in Pakistan by banning flights between East and West Pakistan over its airspace. At a time when the two regions of the country were in conflict, communication between them was decreased by India's act.
40. Peter Woodward, *Sudan, 1898-1989:The Unstable State* (Boulder: Lynne Rienner Publishers, 1990), 223.
41. Orjan Sjoberg and Michael L. Wyzan, "The Balkan States: Struggling Along the Road to the Market from Europe's Periphery," in Orjan Sjoberg and Michael L. Wyzan, eds., *Economic Change in the Balkan States* (New York: St. Martin's Press, 1991) 12.

42. Robert Kaplan, *Balkan Ghosts,* (New York: St. Martin's Press, 1993), xxiv.
43. Sjoberg and Wyzan, p. 2.
44. There was discussion as to how to name the union of Serbia and Montenegro in the absence of the term Yugoslavia. One possibility that was offered with cynical amusement was Crna Serbia, translated as Black Serbia (Montenegro means Black Mountain [Crna Gora], but black also means poor in Serbo-Croatian).
45. See *Nin,* July 10, 1992.
46. David A. Dyker, *Yugoslavia: Socialism, Development and Debt* (London: Routledge, 1990) 183.
47. RFE/RL Research Report, 1, no. 46 (November 20, 1992) 14.
48. *La Repubblica,* September 17, 1992, 15.
49. *The Economist,* June 5, 1993, 57
50. Glenny, 69.
51. Christopher Cviic, *Remaking the Balkans* (New York: Council on Foreign Relations Press, 1991) 43.
52. Sabrina P. Ramet, *Nationalism and Federalism in Yugoslavia,* 2d ed. (Bloomington: Indiana University Press, 1992).
53. Glenny, 88.
54. Nancy Cochrane, "Republic and Provincial Barriers in Yugoslav Agricultural Marketing" presented to the meetings of the American Association for the Advancement of Slavic Studies, Hawaii, 1988.
55. Milovan Djilas, quoted in Kaplan, 75.
56. Among the abundant literature on this subject, see for example, Paul Henze "The Spectre and Implications of Internal Nationalist Dissent: Historical and Functional Comparisons," in *Soviet Nationalities in Strategic Perspective* (London: Croom Helm, 1985); Roman Szporluk, *Communism and Nationalism* (New York: Oxford University Press, 1988); Robert Conquest, ed., *The Last Empire: Nationality and the Soviet Future* (Stanford: Hoover Institution Press, 1986); Nahaylko and Swoboda, *Soviet Disunion* (New York: Free Press, 1989), among others.
57. Nathan Gardels, "Two Concepts of Nationalism: An Interview with Isaiah Berlin," *The New York Review of Books,* November 21, 1991, 19.
58. Conor Cruise O'Brien, "Nationalists and Democrats," *The New York Review of Books,* August 15, 991, 31.
59. Cviic, 29.
60. *The Economist,* October 3, 1992, 55.
61. FRE/RL Daily Report, September 9, 1993.
62. Ethnic mobilization is defined as "potential or actual participation in joint actions when collective membership is based on belonging to the same ethnic group (nationality)," while nationalism is "a political doctrine or social movement which strives to make culture and polity congruent and has the principal aim of

creating a modern nation-state possessing sovereignty over a given territory" (Zaslavsky, 106).

63. In Albania the new nationalists also are making internal demands associated with territory: indeed, it is the nationalist faction of the ruling Democratic Party that are pressuring president Berisha for the restitution of confiscated land. This faction consists mostly of previous landowners and merchants whose wealth was dispersed during world War II, and they have now come to the surface. *The Economist,* June 5, 1993, 57.
64. RFE/RL Daily Report, September 8, 1993, 6.
65. Aleksander Ciric, "Zastrasujuci Pogled Na Druge," *Vreme,* April 29, 1991, 18-20, and quoted in Lenard Cohen, *Broken Bonds* (Boulder: Westview Press, 1993) 239.
66. Zaslavsky, 110.
67. Timothy Garton Ash, "Eastern Europe: The Year of Truth," *The New York Review of Books,* February 15, 1990, 22.
68. *Post-Soviet/East European Report,* 9, no. 38 (17 November, 1992) 3.
69. RFE/RL Research Report, 2, no. 18 (April 30, 1993) 43, 46.
70. Ibid.
71. The author's interviews during the summer of 1993 also show the extent of this nostalgia.
72. *Daily News* (Budapest), July 16-22, 1993.
73. Anders Aslund, "Conclusion: The Socialist Balkan Countries Will Flow East Central Europe," in Sjoberg and Wyzan, 166.
74. This victory occurred without a majority of the vote due to the nature of the election law: indeed, the Croatian Democratic Union (HDZ) had only about 40 percent of the popular vote. New elections in August 1992 again confirmed this phenomenon: the CDU got some 43 percent of the popular vote and ended up with two-thirds of the parliament seats (Robert Hayden, "Constitutional Nationalism in the Formerly Yugoslav Republics," *Slavic Review,* 51, no. 4 [Winter 1992]).
75. In this struggle, the church hierarchy and a Franciscan order are pitted on opposite sides. It has led to an increase in popularity of the Liberal Party (slightly ahead of Tudjman's party in polls in September 1993), which claims the partition of Bosnia is against Croatian interests (FRE/RL Daily Report, June 8, 1993, 7).
76. RFE/RL Research Report, 2, no. 21 (May 21, 1993) 49.
77. The ruling party in response drafted a law in August 1993 to introduce bilingual signs everywhere where a minority is a majority.
78. These nationalist governments share a curious characteristic with some neighboring states—the introduction of intellectuals in the government. Indeed, rarely in history has there been such a proliferation of writers in ruling positions: Vaclav

Havel in Czechoslovakia; Dobrica Cosic and numerous members of the Academy of Sciences in Serbia; Franjo Tudjman, a historian, in Croatia; President Arpad Gonez and Istvan Csurka in Hungary; Ibrahim Rugova, a writer, is leader of Kosovo Albanians, et cetera. One cannot but wonder if their lack of political savvy has something to do with the current political quagmires.

79. Despite assertions that the elections were rigged and that Panic's party did not receive sufficient attention, the overwhelming evidence supports the contention that the population simply did not choose him as the candidate that best responds to their needs. Indeed, he was successful only several urban areas, including Belgrade. He was viewed as too lenient in the negotiations with the Croats, and he is blamed for having made concessions pertaining to land rights on the Prevlaka peninsula and recognition of former Yugoslav republics. In sum, he seemed to be willing to concede much in exchange for the prospect of lifting of sanctions against Yugoslavia in the absence of a commitment by the UN Security Council to do so.
80. Aleksa Djilas "Serbia's Milosevic: A Profile," *Foreign Affairs,* 72, no. 3 (Summer 1993): 93.
81. This symbolism was very important in drawing the distinction between the Croatian ethnic group and others, and it manifested itself in petty symbols such as flags, stamps and street names, all of which had the effect of alienating other groups and straining interethnic relations. This strain was most severely felt in Croat-Serb relations, as Serbs balked at the resurrection of the flag used by the Ustasha regime, stamps with the controversial Cardinal Stepinac on them and the renaming of a city square in the name of Mile Budak, a leader famed for his anti-Semitic actions.
82. Cohen, 276.
83. See chapter 2 for a discussion of the Greek-Macedonian dispute over Macedonian aspirations in Greece.
84. Cohen, 271.
85. RFE/RL Research Report, vol. 2, no. 34 (August 27, 1993) 27.
86. The reasons for Germany's ties to Croatia are many. They include a large Croat population within Germany, which translates into a strong lobby; historical ties, as Croatia was part of Austria-Hungary; cultural ties, as Croats tend to be Germanic speakers (to the extent that they are foreign language speakers); revenge for Serbia's anti-German role in two world wars; and possible access to the Mediterranean, with the inclusion of Croatia into Germany's orbit.
87. Djilas, 96.
88. *The New York Times,* December 22, 1917, 3.
89. William Pfaff, "Invitation to War," *Foreign Affairs,* Summer 1993, 97.
90. *Duga,* July 31, 1993, 6.

Notes to Chapter 2

1. Joseph S. Roucek, *The Politics of the Balkans* (New York: McGraw-Hill, 1939), 5.
2. Quoted in Victor Zaslavsky, "Nationalism and Democratic Transition in Postcommunist Societies" *Daedalus* 121, no. 2 (Spring 1992): 107.
3. Cited in Alfred Pfabigan, "The Political Feasibility of Austro-Marxist Proposals for the Solution of the Nationality Problem of the Danubian Monarchy," in Uri Ra'anan et al., eds., *State and Nation in Multi-Ethnic Societies,* (Manchester: Manchester University Press, 1991), 54.
4. Uri Ra'anan, "Nation and State: Order Out of Chaos" in Uri Ra'anan et al., eds., *State and Nation in Multi-Ethnic Societies* (Manchester: Manchester University Press, 1991), 28.
5. Robert Hayden, "Constitutional Nationalism in the Formerly Yugoslav Republics," paper presented to the conference on "Nation, National Identity and Nationalism" (University of California at Berkeley, September 10-12, 1992), 15.
6. Ronald Steel, *Walter Lippmann and the American Century* (Boston: Little, Brown and Co., 1980), 158.
7. Guglielmo Ferraro, *Tragedija Mira: Od Versaja do Ruha,* (Rome, 1925).
8. *The Economist,* July 13, 1991, 22.
9. Kumar Rupesinghe, ed., *Internal Conflict and Governance* (New York: St. Martin's Press, 1992).
10. Aslavsky, "Nationalism and Democratic Transition in Postcommunist Societies," 98.
11. R. Narroll, "Ethnic Unit Classification," *Current Anthropology* 5, no. 4 (1964).
12. Both of these are personal characteristics, as opposed to the western concept of nationality, which is closely tied to territory and state. Ra'anan, 14.
13. Fredrik Barth, "Introduction," in Fredrik Barth, ed., *Ethnic Groups and Boundaries* (Boston: Little, Brown and Co., 1969), 15.
14. Robert J. Thompson and Joseph R. Rudolph, Jr., "The Ebb and Flow of Ethnoterritorial Politics in the Western World," in Joseph R. Rudolph, Jr., and Robert J. Thompson, eds., *Ethnoterritorial Politics, Policy, and the Western World* (Boulder: Lynne Rienner Publishers, 1989), 2.
15. Frederick L. Shiels, ed., *Ethnic Separatism and World Politics* (Lanham: University Press of America, 1984) and Walter Connor, "Politics of Ethnonationalism," *Journal of International Affairs* 27, no. 1 (1973): 1-21.
16. This study has adopted Smith's definition, according to which nationalism is "a doctrine of autonomy, unity and identity for a group whose members conceive it to be an actual or potential nation" (Anthony Smith, "Chosen Peoples: Why Ethnic Groups Survive," *Ethnic and Racial Studies* 15, no. 3 [July 1992]: 449-450.
17. Grigorij Pomerants, quoted in Zaslavsky, 107.

18. See, for example, Jaroslav Krejci and Vitezslav Velimsky, *Ethnic and Political Nations in Europe* (New York: St. Martin's Press, 1981) among others.
19. Walter Connor, "Nation-Building or Nation-Destroying?" *World Politics* 24 (1972): 320.
20. The countries of Asia and Africa tend to be ethnically heterogeneous, often with compact minorities residing in compact territories. Many of those are divided by international boundaries drawn arbitrarily with little regard to ethnic composition. In Africa alone, the number of ethnic groups divided by international boundaries abound: the Yoruba in Nigeria and Benin; the Hausa in Nigeria, Niger and Ghana; the Berbers in Algeria, Tunisia and Libya; the Bakongo in Angola and Zaire, et cetera. See A. I. Asiwaju, ed., *Partitioned Africans* (New York: St. Martin's Press, 1985). As a result of these imposed unnatural boundaries, Ali Mazrui suggested that the near future will bring about a dramatic redrawing of boundaries across the African continent (*The Economist,* September 11, 1993, 28).
21. Karl Renner (writing under the pseudonym of Rudolf Springer), is quoted in Theodor Hanf, "Reducing Conflict Through Cultural Autonomy: Karl Renner's Contribution," in Uri Ra'anan et al., eds., *State and Nation in Multi-Ethnic Societies* (Manchester: Manchester University Press, 1991) 35.
22. According to Zaslavsky, "It combined divisive measures and integrative techniques to prevent the organization of alliances between neighboring ethnic groups, to undermine the capacity of any nationality to act as a unified entity, and to co-opt the crucial sectors within each nationality into the Soviet regime" (Zaslavsky, 99).
23. One such map is in the *Times Historical Atlas,* edited by Geoffrey Barraclough (Maplewood, New Jersey: Hammond, 1989).
24. One example (and there are far too many) is a recent article by Pfaff in *Foreign Affairs* that claims Vojvodina is largely populated by Hungarians, when in fact the Hungarian population is no greater than 20 percent. It is dangerous when conclusions are drawn upon wrong information. (William Pfaff, "Invitation to War," *Foreign Affairs,* 72, no. 3 [Summer 1993] 99).
25. John Allcock, "Rhetorics of Nationalism in Yugoslav Politics," in John B. Allcock, John J. Horton and Marko Milivojevic, eds., *Yugoslavia in Transition* (New York: Berg, 1992), 283.
26. A team of scholars are trying to prove that Croatian ancestors were Aryans from Persia, and that their name (Hrvat) comes from the ancient word "Hu-Urvat" (an inhabitant of the province Parahvati in the sixth century B.C.).
27. Territorial location of ethnic groups in the former Yugoslavia warrants special attention given the ongoing war and the potential for it to spread to areas of Sandjak, Kosovo and Macedonia. According to the 1991 census, there are very few non-Serb districts in Serbia. The following have Muslim majorities: Novi Pazar (74 percent), Sjenica (75 percent) and Tutin (93 percent); two have Albanian

majorities: Bujanovac (60 percent) and Presevo (90 percent); and two have Bulgarian majorities: Bosilegrad (73 percent) and Dimitrovgrad (52 percent). In Kosovo, the 1991 census takers met with resistance from the Albanian population, and thus only estimates were made: Albanians constitute the majority of the population, 83 percent. In four districts, Serbs are the majority: Zvecan (79 percent), Zubin Potok (53 percent), Leposavic (88 percent) and Stroc (66 percent). In Vojvodina, Serbs are the majority in 35 out of 45 districts, Hungarians in 8 and Slovaks in 2. In Montenegro, 62 percent of the population declared themselves to be Montenegrians, 14 percent Muslims, 9 percent Serbs and 6 percent Albanians. Only three of the districts have a non-Montenegrian majority, two of which have a Muslim majority and one of which has an Albanian majority (*PlanEcon Report,* 8, nos. 14-15 (April 14, 1992): 6).

28. Christopher Cviic, *Remaking the Balkans,* (New York: Council on Foreign Relations Press, 1991), 102. In 1992, Bulgaria conducted a census that included questions on ethnicity, language and religion, thus pitting nationalists against human-rights experts. The results were expected in the fall of 1993.
29. Bogomils are Slavic believers in the Bogomil heresy who converted to Islam when the Turks overran Bosnia in the fifteenth century.
30. Quoted in Peter Berger, "Preface," in Uri Ra'anan et al., eds., *State and Nation in Multi-Ethnic Societies* (Manchester: Manchester University Press, 1991), ix.
31. Smith, 450.
32. For example, in the Balkans, in the nineteenth century, boundary changes did not occur as a result of expansionism by any one power, but rather by the actions of the great powers that traded these territories as if they were pawns.
33. However, the Bulgarian government does not recognize them as such.
34. In the case of Croatia, the World War II years, when the Ustasha regime was in power, are an exception.
35. Quoted in P. Pipinelis, *Europe and the Albanian Question* (Chicago: Argonaut Inc., 1963), 24.
36. This number varies with the sources: 35,000 is from the *Encyclopedia Britannica,* vol. 13, 202. Other sources claim a higher number.
37. See A. J. Toynbee, *A Study of History,* vol. 2 (London: Oxford University Press, 1935) 225.
38. Moreover, the largest population transfer in the Balkans took place at this time, as the Greek and Turkish populations were repatriated to the states of their national origin.
39. Indeed, Croatia benefited greatly in more than territory by aligning itself with the victors in both World War I and II. Indeed, Franklin Roosevelt said on March 15, 1943, that Serbia should have its own government, whereas Croatia should be placed under trusteeship (*Serbia,* no. 7 [January 20, 1992]: 13). That was avoided by the patching up of the country of Yugoslavia.

40. Moreover, there is disagreement over payment for the nuclear power plant in Krsko; the Slovenes claim that the Croats are not paying their bills, leading to an unmet ultimatum issued by the Slovenes and consequent power shortages in Dalmatia.
41. The dispute is over the bed of the River Dragonja, which is the boundary between the two regions. Because a salt works was built near Protoroz, the bed has changed, and there is disagreement over which bed forms the real boundary.
42. Reported by the BBC, cited in FRE/RL Daily Report, May 26, 1993, 5.
43. *Vreme,* April 26, 1993.
44. *La Repubblica,* November 10 and 12, 1992.
45. It is very similar to the questioning of the borders drawn up by the Treaty of Trianon between Czechoslovakia and Hungary: Given that the former no longer exists, should the borders be renegotiated? The issue is further complicated by the fact that Italy and Croatia, and Hungary and Slovakia, were losers of the war, while the Czech Republic and Yugoslavia were winners, and therefore received favorable treatment.
46. It is noted that in the new territorial division of the country, Istria is the only historical region that emerged intact. *Slobodna Dalmacija,* January 16, 1993, 14.
47. Dalmatia and Slavonia were broken into four subunits, and the Serbs were granted several municipalities with a special status still to be determined.
48. RFE/RL Research Report, vol. 2, no. 36 (September 10, 1993): 23.
49. Daniele Joly, *Refugees* (Boulder: Westview Press, 1992), 87.
50. This provoked massive desertions, and fueled the anger of the Austrians, who then went on to exhibit intense cruelty against the Serbian populations. It also represented the first time in this century that the Croats, as part of the Austrian Army, fought against the Serbs.
51. The average proportion of Serbs in Krajina is 62 percent. There is a great variety among the communes, ranging from the following highs and lows: Donji Lapac, 91.1 percent; Vojnic, 88.6 percent; Dvor, 80.9 percent; Pakrac, 38.4 percent; Kostanjica, 55.5 percent and Obrovac, 60.1 percent. (These numbers are taken from the 1981 census, as published in Savezni Zavod Za Statistiku, *Statisticki Godisnjak 1983,* Belgrade, and Jovan Ilic, "Characteristics and Importance of Some Ethno-National and Political-Geographic Factors Relevant for the Possible Political-Legal Disintegration of Yugoslavia," in Stanoje Ivanovic, *The Creation and Changes of the Internal Borders of Yugoslavia* [Belgrade: Srbostampa,1992], Table 6, p. 89).
52. L. S. Stavrianos, *Balkan Federation* (Hamden, Conn.: Ardchon Books, 1964), p. 57
53. Serbs are the majority in 35 out of 45 districts, Hungarians in 8 and Slovaks in 2 (*PlanEcon Report,* vol. 8, nos. 14-15 (April 14, 1992): 6).
54. FRE/FL Research Report, vol. 2, no.15 (April 9, 1993): 44.

55. RFE/RL Research Report, vol. 1, no. 50 (December 18, 1992): 24-25. These demands are extended to Hungarians living in other neighboring states, including Romania and Slovakia. Tensions between the Hungarians and Slovaks on this issue have also increased over the past two years, peaking in the summer of 1993 and leading to a frenzy of governmental and international body meetings (see *Budapest Week,* July 1-7. 1993).
56. FRE/RL Research Report, vol. 2, no. 15 (April 9, 1993): 44.
57. Charles Gati, "From Sarajevo to Sarajevo," *Foreign Affairs* 71, no. 4, (Fall 1992): 68.
58. This has consistently been the goal of the Muslim President Izetbegovic, as he has tried to involve foreign states, especially the United States, into the conflict. Internationalizing the conflict was also partially achieved by the international community's recognizing Bosnia-Herzegovina as an independent state, thereby making any intervention from the Serbs in Serbia or the Croats in Croatia international aggression.
59. It is important to note that the relative size of these three population groups has changed over time. Indeed, according to the census of 1879 carried out by the Austro-Hungarian authorities, the population was 42.8 percent Orthodox Serbs, 38.7 percent Muslims, and 18 percent Catholic Croats.
60. The Muslims, amounting to 43.7 percent of the population, were largely in favor of the Party for Democratic Action (37.8 percent of vote in 1990 elections); the Serbs are 31.3 percent of the population, and the Serbian Democratic Party won 26.5 percent of the votes; the Croats amount to 17.5 percent and their Croatian Democratic Community received 14.7 percent of the votes. Finally, 11.6 percent of the population voted for nonnational parties (Robert Hayden, "The Partition of Bosnia and Herzegovina 1990-1993," RFE/RL Research Report, vol. 2, no. 22 [May 28, 1993]).
61. It was actually much better for the Muslims than the map under discussion in September 1993. They would have received 44 percent of the land, and only 18 percent of their population would have been living outside of the Muslim provinces. The Serbs also would have received 44 percent of the land, but 50 percent of their population would have been outside. Finally, the Croats would have received 12 percent of the land, and 59 percent of their population would have been outside their territory.
62. While there is evidence that the division of Bosnia-Herzegovina among the Serbs and Croats has been under discussion for at least two years, it was publicly stated by both Presidents Tudjman and Milosevic as a possibility only when the new peace plan, reached in Geneva during the summer of 1993, was rejected by the Muslim delegation.
63. Robert Hayden, "The Partition of Bosnia and Herzegovina 1990-1993," 13.

64. An informative book containing statistical evidence on the underlying factors as well as on the manifestations of the war is Srdjan Bogosavljevic, ed., *Bosna i Herzegovina Izmedu Rata i Mira* (Belgrade: Dom Omladine, 1992).
65. Conor Cruise O'Brien, "We Enter Bosnia at Our Peril," *The Independent,* April 23, 1993.
66. Furthermore, scholars such as William Pfaff claims that President Izetbegovic is a true multi-nationalist who attempted to create a multiethnic state (and by implication was prevented from doing so by nationalists of other ethnic/religious groups). (See Pfaff, "Invitation To War"). The fact is that the Bosnian Muslims do not have a good historical track record with respect to their tolerance of other ethnic groups sharing their geographical space. Indeed, their behavior toward the Serbs during Ottoman rule and especially during the World War II indicates an inability and unwillingness to apply the principle of "live and let live." Furthermore, Yugoslavia was a multi-ethnic state: if multiethnicity is a goal to uphold, then why leave the state that represented that goal if not to exert majority rule over other ethnic groups?
67. This region is presently part of the Ukraine, where there are some 135,000 to 200,000 Romanians living (in Chernivtsi Oblast). They have been demanding cultural autonomy and increased relations with Romania. The Ukraine has been relatively lenient with these demands.
68. Alan J. Day, *Border and Territorial Disputes,* 2d ed. (Harlow, Essex: Longman, 1987), 86.
69. FRE/FL Research Report, vol. 2, no. 9 (February 26, 1993): 15.
70. RFE/RL News Brief, vol. 2, no. 7 (1993): 20.
71. The fund only contains $7 million (FRE/RL Daily Report, August 13, 1993, 10).
72. This new republic includes territory that was part of Ukraine. Stalin attached it to the part of Romania that it annexed, in order to dilute the Romanian population. The territory also includes the disputed town of Bender, claimed by both sides. There has also been significant speculation in the press about Romanian military participation in the Moldovan conflict in Dniester, which the Defense Ministry was forced to reject in April 1992 (RFE/RL Research Report, vol. 1, no. 18 (May 1, 1992): 51).
73. As a challenge to Moldovan sovereignty, the Gagauz Supreme Soviet annulled the Romanian language laws in September 1993.
74. *The New York Times,* May 21, 1993.
75. *La Libre Belgique*, February 18, 1993.
76. FRE/RL Research Report, vol. 2, no. 24 (June 11, 1993): 28.
77. The Hungarian organizations have been quite militant in demanding the repeal of ethnic discrimination. Indeed, this party was in part responsible for the creation, in March 1993, of the Council of National Minorities, set up by the Romanian government to address issues pertaining mostly to the Hungarians.

78. The agreement is known as the Carpathian Euro-Region Pact and embraced local cooperation between provinces in Ukraine, Poland, Slovakia and Hungary. In February 1993, it was severely criticized by Romanian President Iliescu.
79. FRE/FL Daily Report, September 13, 1993, 8.
80. Tom Gallagher "Vatra Romanaesca and the Resurgent Nationalism in Romania," *Ethnic and Racial Studies* 15, no. 4 (October 1992): 573.
81. Roucek, 138.
82. FRE/FL Research Report, vol. 2, no. 19 (May 7, 1993): 42.
83. This threat was made by the leaders of the Albanian minority in Macedonia if the Albanians did not receive the same rights as the Macedonians (*The Guardian,* November 17, 1992).
84. FRE/FL Research Report, vol. 2, no. 19 (May 7, 1993): 42.
85. A visitor to Greece in mid-1993 is struck by the overwhelming evidence of the Macedonia issue in Greece. Upon arrival at the Athens airport, one is met by posters proclaiming that "Macedonia is only Greek." A similar statement is available to tourists in tour guides, in banks and so on.
86. Interview in June 1991, *Politika: The International Weekly,* quoted in John Zametica, "The Yugoslav Conflict," *Adelphi,* paper 270 (Summer 1992), 55.
87. FRE/RL Daily Report, September 20, 1993, 8.
88. The official statistical abstract counted 20 percent, but the Albanian sources claim 40 percent.
89. RFE/RL Research Report, vol. 1, no. 27 (November 27, 1992): 33.
90. Zametica, 30.
91. Sabrina Ramet, "War in the Balkans," *Foreign Affairs,* 83; Zametica, 25.
92. Aleksa Djilas, "Serbia's Milosevic," *Foreign Affairs,* Summer 1993, 82. Most of these migrations, he claims, were forced by the Albanian extremists.
93. Zametica, 26.
94. While the majority of Albanian writers welcome the idea of a unification of all Albanian-inhabited lands, some question it on grounds that the Yugoslav Albanians and the Albanian Albanians have had such divergent histories, lived under such different economic systems and experienced such different levels of development that a union would introduce conflict into a society.
95. *Borba,* May 22, 1993.
96. RFE/RL vol. 12, no. 46 (November 20, 1992): 17.
97. *Post-Soviet/East European Report,* vol. 9, no. 37 (November 3, 1992): 5.
98. *Dubrovacki Vjesnik,* September 23, 1993, 4.
99. RFE/RL Daily Report, October 7, 1993, 6.
100. RFE/RL Daily Report, October 26, 1993, 6.
101. *Athens News,* July 4 and 5, 1993.

102. The United Nations has been pressing ahead on negotiations in Cyprus, in part to avoid comparisons with the standstill in Croatia. *Athens News,* July 8, 1993.
103. Pipinelis, 67. Another census was taken by the Turks in 1909, according to which 128,050 of a population of 223,611 were Greek Orthodox. This estimate has been claimed to be an underestimate, given the desire of the Turkish census takers to augment the size of the Muslim population. Henry Baerlein, *Southern Albania* (Chicago: Argonaut Inc., 1968), 7.
104. RFE/RL Research Report, vol. 2, no. 33 (August 20, 1993): 32.
105. One such figure is the Archbishop Sevastionos in Konitsa in Epirus (Cviic, 101)
106. RFE/RL Research Report, vol. 2, no. 33 (August 20, 1993): 31.
107. *Balkan News,* July 11, 1993, 6.
108. Part of the reason that the Mitsotakis government has responded so strongly to the expulsion of the Orthodox cleric is that the opposition party, PASOK, has made an issue of it, and Prime Minister Mitsotakis wanted to show strength on the issue. I was in Greece at the time of the expulsion, and it was certainly the principal topic of debate in political circles as well as in the media.
109. Zametica, 23.
110. Alfred Serreqi, quoted in *Post-Soviet/East European Report,* vol. 9, no. 39 (December 1, 1993): 3.
111. He said, "When I gave utterance to these words ['that all nations had a right to self-determination'] I said them without the knowledge that nationalities existed, which are coming to us day after day. . . .You do not know and cannot appreciate the anxieties that I have experienced as the result of the many millions of people having their hopes raised by what I have said." Quoted in Stanley Kober, "Revolutions Gone Bad," *Foreign Policy* 91 (Summer 1993): 68.
112. Quoted in Pfaff, 100.
113. Amitai Etzioni, "The Evils of Self-Determination," *Foreign Policy* 89 (Winter 1992-93): 33.
114. *La Repubblica* June 29, 1993.
115. *The Economist,* July 31, 1993, 45. With respect to the Slovak constitution and its treatment of minorities, see RFE/RL vol. 1, no. 43 (October 30, 1992): 39-42.
116. Robert Hayden, "Constitutional Nationalism in the Formerly Yugoslav Republics" *Slavic Review,* 51, no. 4 (Winter 1992).
117. Hayden, 4.
118. RFE/RL vol. 1, no. 46 (November 20, 1992): 16.
119. Oszkar Jaszi (1926), cited in Peter Sipos, "National Conflicts and the Democratic Alternative in the Austro Hungarian Monarchy and Its Successors," in Uri Ra'anan et al., eds., *State and Nation in Multi-Ethnic Societies* (Manchester: Manchester University Press, 1991), 105.

120. Indeed, even Secretary General Boutros-Ghali announced that the UN will not recognize the results, despite the request by the Bosnian Serbs that the UN monitor the balloting (RFE/RL Daily Report, May 12, 1993, 8).
121. James Simmie and Joze Dekleva, eds., *Yugoslavia in Turmoil: After Self-Management?* (London: Pinter Publishers, 1991) xviii.
122. David A. Dyker, *Yugoslavia: Socialism, Development and Debt* (London: Routledge, 1990), 184.
123. This was revealed to UN representative Cyrus Vance (Lenard J. Cohen, *Broken Bonds,* (Boulder: Westview Press, 1993), 235.
124. Roucek, chapter 1.

Notes to Chapter 3

1. *The Economist,* September 11, 1993, 57.
2. Dennis Rusinow, "Yugoslavia: Balkan Breakup?" *Foreign Policy* 83 (Summer 1991).
3. Andrew Freris, *The Greek Economy in the Twentieth Century* (New York: St. Martin's Press, 1986), ch. 5.
4. Christopher Cviic, *Remaking the Balkans* (New York: Council on Foreign Relations, 1991), 48.
5. John R. Lampe and Marvin R. Jackson, *Balkan Economic History 1550-1950* (Bloomington: Indiana University Press, 1982); Nicholas Giannaris, *The Economies of the Balkan Countries* (New York: Praeger, 1982); George Hoffman, *Regional Development Strategy in Southeast Europe: A Comparative Analysis of Albania, Bulgaria, Greece, Romania and Yugoslavia* (New York: Praeger 1972); Orjan Sjoberg and Michael L. Wyzan, *Economic Change in the Balkan States* (New York: St. Martin's Press, 1991).
6. Lampe and Jackson, 594
7. *The Economist,* May 22, 1993, 6 (survey).
8. Mark Mazower, *Greece and the Inter-War Economic Crisis* (Oxford: Clarendon Press, 1991), 237.
9. RFE/RL vol. 1, no. 41 (October 16, 1992): 49.
10. RFE/RL vol. 1, no. 43 (October 30, 1992): 57.
11. The Bank of Slovenia, cited in "Key Facts: Republic of Slovenia," *Financial Times* Survey, March 30, 1993.
12. The data are from PlanEcon, printed in *The Economist,* June 26, 1993, 55.
13. RFE/RL Research Report, vol. 2, no. 3 (January 15, 1993): 34.
14. *Transition,* vol. 4, no. 2 (March 1993): 12.
15. *PlanEcon Report,* vol. 8, nos. 14-15 (April 14, 1992): 2.
16. *Transition,* vol. 4, no. 2 (March 1993): 12.

17. *The Economist,* September 4, 1993, 6.
18. RFE/RL Research Report, vol. 2, no. 4 (January 22, 1993): 45.
19. FRE/RL Research Report, vol. 2, no. 23 (June 4, 1993): 27.
20. Croatian Economic Trends, no. 3, 1993, cited in REF/RL Research Report, vol. 2, no. 26 (June 25, 1993): 33.
21. RFE/RL Research Report, vol. 2, no. 34 (August 27, 1993): 27.
22. Interview with Tomislav Popovic, printed in *Duga,* July 31, 1993, 11.
23. RFE/RL Research Report, vol. 2, no. 23 (June 4, 1993): 28.
24. *The Economist,* August 14, 1993, 55.
25. RFE/RL News Brief, vol. 2, no. 7 (February 1-5, 1993): 13.
26. FRE/RL News Brief, vol. 2, no. 11 (March 1-5, 1993): 11.
27. FRE/RL Daily Report, August 31, 1993, 7.
28. RFE/RL Daily Report, May 28, 1993, 8.
29. *Transition,* vol. 4, no. 2 (March 1993): 12.
30. RFE/RL Research Report, vol. 2, no. 3 (January 1, 1993): 117.
31. RFE/RL Daily Report, vol. 2, no. 36 (August 30-September 3, 1993): 12.
32. RFE/RL Research Report, vol. 1, no. 41 (October 16, 1992): 49.
33. RFE/FL Research Report, vol. 1, no. 42 (October 23, 1992): 38.
34. RFE/RL Research Bulletin, vol. 10, no. 1 (January 5, 1993): 6.
35. RFE/RL Research Report, vol. 2, no. 32 (August 13, 1993): 55.
36. Macedonian Information and Liaison Service, *Dnevni Vesti Angliski,* April 30, 1993.
37. RFE/RL Research Report, vol. 1, no. 22 (May 29, 1992): 40.
38. Both public and private sector workers paralyzed Greece in August in protest over austerity measures of the Greek government. Due to strikes by the state electrical company, energy rationing occurred; banks closed; the strike of drivers and gas station attendants forced the mobilization of the army to provide transportation for the one-third of Greeks that live in Athens and its suburbs.
39. RFE/RL Research Report, vol. 1, no. 42 (October 23, 1992): 38.
40. RFE/RL Research Bulletin, vol. 10, no. 1 (January 5, 1993): 1, 5.
41. RFE/RL Research Report, vol. 1, no. 43 (October 30, 1992): 57.
42. RFE/RL Research Bulletin, vol. 10, no. 1 (January 5, 1993): 6.
43. RFE/RL Research Report, vol. 1, no. 41 (October 16, 1992): 49.
44. RFE/RL Research Report, vol. 2, no. 9 (February 26, 1993).
45. It was originally to be called the *kruna,* but in order to appease the political right, it was changed to the *kuna,* the same name it carried during the Ustasha government of World War II.
46. RFE/RL Research Report, vol. 2, no. 3 (January 15, 1993): 35.
47. FRE/RL Research Report, vol 2, no. 32 (June 4, 1993): 27.
48. See Boris Pleskovic and Jeffrey Sachs, "Currency Reform in Slovenia: The Tolar Standing Tall," in *Transition,* vol. 3, no. 8 (September 1992).

49. This prognosis was stated by *The Economist,* September 4, 1993, 6.
50. FRE/RL Daily Report, August 31, 1993, 7. However, see *Vreme* (August 9, 1993, p. 12) for variation on this number, depending on how inflation is measured.
51. RFE/RL News Brief, vol. 2, no. 11 (March 1-5, 1993): 13.
52. The front page of *Vreme,* the opposition weekly, reads "Dinari Odlaze, Dinaroidi Dolaze" ("The dinars are leaving, the dinaroids are coming," implying size), *Vreme Novca,* 19 (July 26, 1993).
53. For Romania: RFE/RL Research Report, vol. 1, no. 43 (October 30, 1992): 57; for Slovenia: *Financial Times.* According to the Slovenian central bank, the surplus follows a deficit in 1991 of $257 million.
54. RFE/RL Research Report, vol. 2, no. 1 (January 1, 1993): 118.
55. *PlanEcon Report,* vol. 8, nos. 14-15 (April 14, 1992): 3.
56. This is Macedonia's largest success to date in establishing new markets to replace the lost Yugoslav ones: the order to manufacture buses for Turkey is worth more than $100 million.
57. In the case of Macedonia, disintegration robbed enterprises of their traditional markets, especially in the food, textile and tobacco industries.
58. According to *Transition,* some 70 percent was with the West.
59. M. Grubisic, "Inter-republican Trade in Goods and Services 1967-1987," *Analize i Prikazi* 73 (1990).
60. Vesna Bertoncelj-Popit, "Markovic Convoy," *Slovenian Business Report* 1 (September 1991): 4.
61. See Table 3.2 below.

Table 3.2
Yugoslav Interregional Trade in 1987 (percent)

	Inflows from republic markets	Inflows from foreign markets	Outflows to republic markets	Outflows to foreign markets
Yugoslavia	—	9.6	—	10.4
Bosnia-Herzegovina	60.4	8.2	69.4	9.5
Montenegro	39.2	8.9	60.0	10.5
Croatia	64.8	9.8	68.6	10.8
Macedonia	59.5	11.7	66.5	9.2
Slovenia	61.9	12.1	62.9	13.5
Serbia	71.9	8.5	76.2	8.7

Source: Andra Milojcic, *Vrednost Nabavki i Isporuka Izmedu Republika i Pokrajina u 1987 Godini,* Saopstenje, no. 324 (November 1991).

62. *Vreme*, October 19, 1992, 33.
63. *Transition,* vol. 2, no. 2 (February 1991): 7.
64. Alan H. Smith, "Change in East Central Europe: Effect on the Balkan Economies," in Orjan Sjoberg and Michael L. Wyzan, *Economic Change in the Balkan States* (New York: St. Martin's Press, 1991), 158.
65. FRE/RL News Brief, March 29-April 2, 1993, 12.
66. *The Wall Street Journal,* December 7, 1992.
67. FRE/RL Research Report, vol. 2, no. 32 (August 13, 1993): 55.
68. Ibid., 54.
69. RFE/RL News Brief, vol. 2, no. 3 (1993, December 28-January 8, 1993): 22.
70. Cviic, 61.
71. The aid is mostly for humanitarian purposes, largely for food, and started as Operation Pelican from Italy. By 1992, the EC took over as principal aid donor. However, Turkey is also very active in Albania. Not only does it provide humanitarian aid, which amounted to $21.9 million (in 1991-1992), but it is also highly involved in Albania's economic restructuring. This included the joint transportation venture Alb-Balkan Interbut, the granting of scholarships—both civilian and military—for education in Turkey and the setting up of a Turkish school in Tirana. But most importantly, Turkey's policy is one of long-term cooperation: when Turkish president Ozal visited Albania in 1993, he offered it a 15-year economic program to renovate its ports, develop its tourism and develop its financial institutions and military cooperation (RFE/RL Research Report, vol. 2, no. 11 (March 12, 1993): 32).
72. National Statistical Service of Greece, *Statistical Yearbook of Greece 1989* (Athens, 1992), table 23, p. 508.
73. *The New York Times,* May 30, 1993.
74. *Croatian Economic Trends,* nos. 1-2 (1993): 16.
75. Thomas Poulsen, "Yugoslavia in Geographical Perspective" in John Allcock et al., eds., *Yugoslavia in Transition* (New York: Berg, 1992), 61.
76. Interview published in *Balkan News,* July 18, 1993.
77. RFE/RL Daily Report, September 28, 1993, 9.
78. RFE/RL Research Report, vol. 2, no. 26 (June 25, 1993): 34. Military expenditures are not listed as a separate item in the budget, so this represents a lower limit, since many of the weapons purchases are in defiance of the embargo and thus will be outside of the state budget.
79. *The Economist,* May 22, 1993, 14 (survey).
80. *The Economist,* June 26, 1993, 55.
81. Jose Mencinger cited in "Republic of Slovenia," *Financial Times* Survey, March 30, 1993.

82. Ljubisa Adamovic, "The Position and Strategy of Serbia in the New European Order," in Tomislav Popovic, ed., *The Position and Strategy of Serbia in the New European Order* (Belgrade: Institute of Economic Sciences, 1992), 88.

Notes to Chapter 4

1. James Simmie and Joze Dekleva, eds., *Yugoslavia in Turmoil: After Self-management?* (London: Pinter Publishers, 1991), xvi.
2. This analogy was quoted in *China Daily,* and reprinted in *Transition,* vol. 3, no. 8 (September 1992): 9.
3. Charles Gati, "From Sarajevo to Sarajevo," *Foreign Affairs* 71, no. 4 (Fall 1992): 64.
4. Michel Piessel, *The Secret War in Tibet* (Boston: Little, Brown, 1972), 126.
5. Milica Zarkovic Bookman, *The Economics of Secession* (New York: St. Martin's Press, 1993).
6. This view is embodied in Francis Fukuyama, *The End of History and the Last Man* (New York: The Free Press, 1992).
7. This organization of the elements of transition to a market economy is offered by Susan M. Collins and Dani Rodrik, *Eastern Europe and the Soviet Union in the World Economy* (Washington: Institute for International Economics, May 1991), 11. This paragraph draws heavily on their analysis.
8. Jeffrey Sachs, "Poland and Eastern Europe: What Is to Be Done?" in Andras Koves and Paul Marer, eds., *Foreign Economic Liberalization* (Boulder: Westview Press, 1991).
9. Karoly Attila Soos, "Liberalization and Stabilization," in Andras Koves and Paul Marer, eds., *Foreign Economic Liberalization* (Boulder: Westview Press, 1991); Jude Wanniski, "IMF's Economic Massacre in the Balkans," *The Wall Street Journal,* August 10, 1993.
10. These slower changes are associated with Hungary. Poland began reforms with a Big Bang; however, by early 1992, it had lost its nerve and adopted Hungary's more gradualist approach of privatization.
11. The response to this view was given by Al Hirshman, namely that if a country could develop all sectors simultaneously, there would be no problem of economic development. A similar response might have been given to Big Bangers.
12. *Transition,* vol. 3, no. 5 (May 1992): 8.
13. Before the demise of communism, Dubchek and Markovic were popular in the West for their views which were perceived as anticommunist. Actually, these people were Marxists in search of a humane socialism to be applied to their societies. After the demise of communism, they remained true to their ideas, but

the world changed, and they were no longer seen as opponents of communism but rather the embodiment of communism itself. They are countered by Janos Kornai, who said that a third way between communism and capitalism is impossible (see *The Economist,* April 24, 1993, 54). Czurska rejects both capitalism and communism as models for Hungary, and opts instead for the "third road" (see RFE/RL Research Report, vol. 1, no. 40 (October 9, 1993): 28.

14. Although some, like Sean Gabb, view the costs of the application of the third way in Slovakia as significantly lower than those of the increased Germanization of the Czech Republic ("Czechs, not Slovaks, Headed for Trouble," *The Prague Post,* September 1-7, 1992, 15).
15. *The New York Times,* October 19, 1992. Some loosening was evident in mid-1993, as China attempted to win a bid to host the Olympic games in the year 2000.
16. As described by Mark Kramer, "the decline in Eastern Europe's industrial production between 1990 and 1992 came exclusively in the inefficient state sector, where most of what was 'lost' would have had no place in a viable free-market economy anywhere. The production statistics understated the growth and vibrancy of the burgeoning private sector in the region, especially the rise of small-scale entrepreneurs."
17. Charles Gati, "From Sarajevo to Sarajevo," *Foreign Affairs* 71, no. 4, (Fall 1992): 74.
18. *The Economist,* October 3, 1992, 19.
19. RFE/RL News Brief, vol. 2, no. 19, 19.
20. RFE/FL Research Report, vol. 2, no. 1 (January 1, 1993): 110.
21. Ibid.
22. RFE/RL Research Report, vol. 2, no. 32 (August 13, 1993): 54.
23. RFE/RL Research Report, vol. 1, no. 46 (November 20, 1992): 36-37.
24. FRE/RL Research Report, vol. 2, no. 27 (July 2, 1993).
25. The IMF is dissatisfied with Romania's exchange and interest rates and the government's efforts to rectify the situation.
26. Kjell Engelbrekt, "Bulgaria: The Weakening of Postcommunist Illusions," FRE/FL Research Report, vol. 2, no. 1 (January 1, 1993): 80.
27. The sale of state property will not include mining and power, oil processing, rail transport or military production industries.
28. RFE/RL Research Report, vol. 2, no. 32 (August 13, 1993): 55.
29. The director of the Agency for Privatization complained that politicians were unwilling to act on privatization issues (RFE/RL News Briefs, vol. 2, no. 17 (1993): 15).
30. FRE/FL Research Report, vol. 2, no. 21 (May 21, 1993): 37.
31. This decree, issued in July 1990, contained guarantees against expropriation or nationalization and contained assurances pertaining to repatriation of profits. It represented a sharp break with the past, during which a policy of self-reliance

(following the break with China in 1978) restricted contacts with the world economy.

32. Elez Biberaj, "Albania's Bumpy Road to the Market," in *Transition,* vol. 2, no. 2 (February 1991): 9.
33. Critics of the Papandreou policies claim that state intervention was pervasive in all areas of economic activity: the public sector accounts for more than 70 percent of all economic activity. The fiscal deficit was exploding, the public debt rising at an alarming rate, wage and price controls were in operation, nationalizations were widespread, the civil service was in near disintegration and public investment in infrastructure had been neglected in favor of consumption. In the 1980s, social security transfers rose to 14 percent of GDP, up from 6.5 percent in the 1960s. Over the same period, current government disbursements as a percentage of GDP went up by 20 percent points, to 41 percent of GDP.
34. Some 50 firms are in the process of being privatized. Although the number is small, their importance in the economy is large: for example, in the cement industry, they account for 45 percent of total production and 50 percent of exports (*The New York Times,* November 11, 1992).
35. Yugoslavia established a 60-day limit on arrears, and that coupled with a strict monetary constraint resulted in a surge of enterprise failures that meant an increase in unemployment, or the impoverization of the workers as enterprises missed wage payments to avoid bankruptcy.
36. Zivko Pregl, "Programme of Reforms in Yugoslavia," in James Simmie and Joze Dekleva, eds., *Yugoslavia in Turmoil: After Self-Management?* (London: Pinter Publishers, 1991).
37. Simmie and Dekleva, xvi.
38. *Euromoney Supplement,* May 1992: 7.
39. Many place much hope in the Croatian diaspora. As it played a large role in the war effort, so too they are expected to be involved in investment.
40. This claim is made in Boris Pleskovic and Jeffrey Sachs, "Currency Reform in Slovenia: The Tolar Standing Tall," in *Transition,* vol. 3, no. 8 (September 1992): 6-8. The section on Slovenian transition draws heavily on this article.
41. Privatization has taken place less quickly. Indeed, parliamentary discussion on privatization had been repeatedly postponed as of May 1992. See Kenneth Zapp, "Slovenia: A Case Study of the Challenges of Privatization," paper presented to the Biennial Conference on Eastern Europe, Sarasota, Florida, March 1993, and Ivo Bicanic, "Privatization in Yugoslavia's Successor States," RFE/RL Research Report, vol. 1, no. 22 (May 29, 1992): 45-46.
42. During the month of conversion to the tolar (October 1991), inflation peaked at 21.5 percent per month. Since then, it fell to 5.1 percent in April 1992 and 1.2 percent in August 1992 (Pleskovic and Sachs, 7).

43. RFE/RL Research Report, vol. 2, no. 3 (January 15, 1993): 34.
44. Discounts were offered to those purchasing in foreign currency.
45. These are described in an unpublished document: Federal Government of Yugoslavia, *Program for the Macroeconomic Stabilization of Yugoslavia* (Belgrade, October 1992).
46. *Vreme News Digest Agency,* no. 87 (May 24, 1993).
47. Gordon H. McCormick and Richard E. Bissell, *Strategic Dimensions of Economic Behavior* (New York: Praeger 1984).
48. Michael Mann, *States, War and Capitalism: Studies in Political Sociology* (Oxford: Basil Blackwell, 1988).
49. Henry Barbera, *Rich Nations and Poor in Peace and War* (Lexington: Lexington Books, 1973): 3
50. These are derived from Alfred C. Neal, ed., *Introduction to War Economics* (Chicago: Richard D. Irwin, 1942).
51. I. Bicanic claims that this is the estimated material costs of the war in Croatia, outside of Krajina (RFE/RL Research Report, vol. 2, no. 26 [June 25, 1993]: 34).
52. Given the galloping inflation rate, purchase of private property in foreign currency has become ridiculously cheap. For example, housing of some 70 square meters, previously socially owned, could be purchased for some 30 deutsche marks in mid-1993.
53. *Interviju,* August 6, 1993, 16.
54. *Euromoney* Supplement, May 1992, 2.
55. RFE/RL News Brief, December 28-January 8, 1993, 14.
56. *Vjesnik,* April 6, 1992.
57. *Euromoney,* May 1992 Supplement, 8.
58. RFE/RL Research Report, vol. 2, no. 26 (June 25, 1993): 34.
59. FRE/RL Daily News, June 16, 1993.
60. Sabrina P. Ramet, "War in the Balkans," *Foreign Affairs* 71, no. 4 (Fall 1992): 91.
61. This was done in March 1992, and at that time, the foreign debt was on the order of \$2.5 billion. Meanwhile the regional responsibility for the payment of the former Yugoslav debt was to be assessed, since it did not seem logical that Serbia should carry the entire weight of the former Yugoslav debt burden.
62. There is some controversy over this point, as some scholars believe that the war measures taken by the Bolsheviks were in fact part of an ideological package that would have been enacted even in the absence of war. See Paul R. Gregory and Robert C. Stuart, *Soviet Economic Structure and Performance,* 4th ed. (New York: Harper and Row Publishers, 1990), 51.
63. See Tomislav Popovic, "Alternativni Scenariji Uredjenja Odnosa Sa Medjunarodnim Okruzenjem i Ocene Njihovih (De)stabilizacionih Efekata," in *Osnove Stabilizacionog Programa* (Belgrade, Institut Ekonomskih Nauka, 1992), and

Branko Hinic and Rajko Bukvic, "Efekti Mera Ekonomske Politike i Ocekivana Privredna Kretanja u Narednom Periodu," unpublished paper, Belgrade 1992.

64. Popovic, 4.
65. *The New York Times,* October 1, 1992.
66. Gati, 74.
67. *The Wall Street Journal,* December 7, 1992.
68. The poll was conducted for *The Wall Street Journal Europe* and the German newspaper *Handelsblatt* (*The Wall Street Journal Europe,* July 8, 1993).

Notes for Chapter 5

1. Quoted in Gary Clyde Hufbauer, Jeffrey J. Schott and Kimberly Ann Elliot, *Economic Sanctions Reconsidered* (Washington: Institute for International Economics, 1990), 9.
2. Hufbauer, Schott and Elliot, 3.
3. Misha Glenny, *The Fall of Yugoslavia* (London: Penguin, 1992), 103.
4. Hufbauer, Schott and Elliot, 40.
5. Ibid., 3.
6. It is interesting that all but three former Soviet states voted for this expulsion, as did all East European states.
7. RFE/RL News Brief, vol. 2, no. 23 (May 24-28, 1993): 19.
8. RFE/RL New Brief, vol. 2, no. 26 (June 14-18, 1993): 10.
9. For example, the purchase of a subcompact automobile required over ten years' salary in mid-1993, while it only required eight months' salary in 1990.
10. *Duga,* September 25, 1993, 28.
11. FRE/RL Daily Report, May 28, 1993, 8.
12. Branko Hinic and Fajko Bukovic, "Efekti Mera Ekonomske Politike i Ocekivana Provredna Kretanja U Narednom Periodu," unpublished paper, Belgrade, 1992.
13. *Vreme News Digest Agency,* no. 88 (May 31, 1993).
14. *Vreme,* May 3, 1993.
15. *Vreme,* May 3, 1993.
16. *Duga,* September, 1993.
17. Sabrina Petra Ramet, "War in the Balkans," *Foreign Affairs* 71, no. 4 (Fall 1992): 90.
18. *Danas,* March 12, 1993.
19. *The Miami Herald,* August 26, 1992.
20. RFE/RL Daily Report, September 15, 1993, 6.
21. *Tanjug,* August 24, 1993.
22. Another social side effect of the sanctions and the concomitant decrease in the standard of living is the decrease in the divorce rate in urban Serbia. People cannot

afford to separate and thus are staying together for economic reasons (*Politika,* July 25, 1993).

23. *The New York Times,* June 26, 1992.
24. *The Economist,* October 9, 1993, 60.
25. Dimitrije Boarov, the economics writer for *Vreme,* quoted in the *Miami Herald,* August 26, 1992.
26. RFE/RL Research Report, vol. 2, no. 21 (May 21, 1993): 50.
27. Hinic and Bukovic, see especially figures 1 and 3.
28. RFE/RL Research Report, vol. 2, no. 34 (August 27, 1993): 21.
29. Sanctions were not really expected right up until the end. This is evident by the fact that a clear plan for the measures to be taken was not published until September 1992, some three months after the fact (Privredna Komora Srbije, *Predlog Mera Za Odrzavanje Proizvodnje u Uslovima Ekonomske Blokade,* Belgrade, September 4, 1992).
30. Indeed, in South Africa, this period was a couple of years, while in the case of Italy, it was short (Margaret P. Doxey, *Economic Sanctions and International Enforcement* [London: Oxford University Press, 1991]).
31. *The Wall Street Journal,* October 7, 1992.
32. *The Miami Herald,* August 26. 1992.
33. RFE/RL Research Report, vol. 1, no. 26 (June 26, 1992): 36.
34. *The New York Times,* December 6, 1992.
35. *Politika* July 23, 1993.
36. RFE/RL Daily Report, August 25, 1993: 6.
37. *Interviju*, August 6, 1993.
38. Tomislav Popovic, "Alternativni Scenariji Uredjenja Odnosa Sa Medjunarodnim Okruzenjem i Ocene Njihovih De-Stabilizacionih Efekata," in *Osnove Stabilizacionog Programa* (Belgrade: Institut Ekonomskih Nauka, 1992).
39. *UPI Reports,* August 13, 1992.
40. *Vreme,* May 3, 1993.
41. *The Wall Street Journal,* October 7, 1993. The fact that it still looks devastated is the result of a reluctance on the part of the inhabitants to rebuild until its final status is resolved. Reconstruction is not likely while the resumption of war remains a possibility.
42. *The New York Times,* August 31, 1992.
43. *The Miami Herald,* August 26, 1992.
44. RFE/RL Daily Report, October 14, 1993, 8.
45. This sentiment is expressed in *Interviju,* November 27, 1992: 16-17.
46. *Inat* is a word is a word that is difficult to translate into English. More than a word, it represents a mentality. It can roughly be translated as "spite."

47. Indeed, having crossed the border between Hungary and Yugoslavia on several occasions during the summer of 1993, I was surprised to find that there was never serious inspection, let along the presence of UN monitoring personnel.
48. RFE/RL Research Report, vol. 1, no. 26 (June 26, 1992): 36.
49. Ibid., 36.
50. Radio Bucharest, June 4, 1992, quoted in ibid., 37.
51. Ibid., 34.
52. *UPI Reports,* August 18, 1992.
53. *Pravda,* December 15, 1992. This estimate includes the losses from Libya and Iraq as well as Yugoslavia.
54. FRE/FL Daily Report, May 18, 1993, 9.
55. An interview with President Zhelyu Zhelev in the *Neue Zuercher Zeiturn,* February 22, 1993.
56. RFE/RL Research Report, vol. 2, no. 30 (July 23, 1993): 40.
57. RFE/RL Daily Report, September 15, 1993,7.
58. Interview with Kiro Gligorov in *Interviju,* November 27, 1992, 10.
59. *Serbian Unity Congress Weekly Bulletin* 27, September 3, 1993.
60. *Slobodna Dalmacija,* January 16, 1993, 13.
61. These measures were suggested in a declaration adopted by the Bulgarian government and addressed to the 48th session of the UN General Assembly in September 1993.
62. RFE/RL News Brief, June 14-18, 1993, 10.
63. Hufbauer, Schott and Elliot, 43-55.
64. The no-fly zone continues to be monitored, at great expense to western governments, in an effort to curb Serbian incursions over Bosnian airspace. This seems all the more ludicrous given the evidence presented by the United Nations that since October 1992, the vast majority of incursions were not by Serbian aircraft.
65. The threat of U.S. bombers escalated in February 1993 and again in the summer of 1993. However, the second time around, the threat was taken less seriously in Serbia and Montenegro, due in large part to what was perceived as wavering and lack of understanding of the situation by President Clinton. Indeed, the joke making the rounds at the time was: "President Clinton has just announced in Washington that if the Serbs do not immediately withdraw from Mt. Ingman [outside Sarajevo] he was immediately going to bomb Saddam Hussein!"
66. *The New York Times,* May 27, 1992.
67. Hufbauer, Schott and Elliot, 73.
68. Institut Ekonomskih Nauka, *Osnove Stabilizacionog Programa, Sa Epilogom,* Belgrade, 1993, Draft Document, 69-71.
69. Andra Milojic, *Vrednost Nabavki i Isporuka Izmedu Republika i Pokrajina u 1987 godini,* Saopstenje, 324 (November 1991).

70. This incident clearly showed that the government in Belgrade does not set the rules for the Serbian government in Pale, where the Vance-Owen Peace Plan was overwhelmingly rejected. It is also noted that it is not sanctions directly that have effected this change of heart, but rather the indirect way in which sanctions have caused Montenegro to reevaluate its historical ties to Serbia. Indeed, there is evidence that President Bulatovic of Montenegro became unwilling to continue supporting sanctions and has threatened to leave the Yugoslav federation if such an action would result in the immediate lifting of sanctions against Montenegro. This is the real pressure that forced Milosevic's hand
71. Especially in Iraq, where sanctions were aided by military incursions under the auspices of the United Nations to achieve at least part of their political goal.
72. Victor Zaslavsky, "Nationalism and Democratic Transition in Postcommunist Societies," *Daedalus* 121, no. 2 (Spring 1992), 116.
73. *Politika,* July 22, 1993.
74. RFE/RL Research Report, vol. 2, no. 21 (May 21, 1993): 54.
75. Interview with Slobodan Jarcevic, *Duga,* September 25, 1993: 6.
76. *Vreme News Digest Agency,* no. 88, May 31, 1993.
77. Citations from *The Observer* are published in *Vreme,* August 9, 1993, 15.

Notes to Chapter 6

1. *U.S. News and World Report,* November 30, 1992, 36.
2. *La Repubblica,* August 20, 1992.
3. *The International Herald Tribune,* July 14, 1993.
4. RFE/RL Daily Report, May 14, 1993.
5. RFE/RL Daily Report, June 11, 1993, 6. Of these, 500 have been relocated to Pakistan as guests of the government in Islamabad (RFE/RL Daily Report, June 17, 1993, 6).
6. Leon Gordenker, *Refugees in International Politics* (New York: Columbia University Press, 1987), 63.
7. Gordenker, 52-59.
8. An excellent study on migration in the Balkans is Radovan Samardzic and Dimitrije Dordevic, *Migrations in Balkan History* (Belgrade: Institute for Balkan Studies Special Edition no. 39, 1989).
9. Geoffrey Barraclough, ed., *The Times Atlas of World History,* 3d ed. (Maplewood, New Jersey: Hammond, 1989), 265.
10. Alex N. Dragnich and Slavko Todorovich, *The Saga of Kosovo* (Boulder: East European Monographs, 1984), 158. There is evidence that Serbs were prevented from returning by President Tito, thus altering the demographic balance in Kosovo.

11. Aleksa Djilas, "Serbia's Milosevic," *Foreign Affairs* 72, no. 3 (Summer 1993): 82.
12. The Turkish government claimed that by mid-October 1992, Turkey had received 160,000 ethnic Turks from Bulgaria, a number hotly disputed by the Bulgarian authorities. This was claimed by Turkish Prime Minister Suleyman Demirel in *Otechestven Vestnik,* October 12, 1992, and cited in RFE/RL Research Report, vol. 1, no. 45 (November 13, 1992), 1.
13. Daniele Joly, *Refugees-Asylum in Europe* (Boulder: Westview Press, 1992), 75. and FRE/RL Research Report, vol. 2, no. 18 (April 30, 1993).
14. *UPI Reports,* February 5, 1993.
15. Joly, 76-77.
16. This number was given by Ivan Dundarov in an interview on March 12, 1992. Quoted in Joly, 77.
17. FRE/RL Research Report, vol. 1, no. 47 (November 27, 1992): 64.
18. Joly, 79.
19. FRE/RL Research Report, vol. 2, no. 3 (January 15, 1993), 1.
20. *Vecernji List,* January 16, 1993, 5.
21. *The New York Times,* November 28, 1992.
22. RFL/RL Daily Report, May 11, 1993.
23. *The New York Times,* October 14, 1992.
24. This ruling exempts some categories, such as mothers of young children, orphans, and the sick or wounded (*The New York Times,* June 16, 1993).
25. Mesic identified three stages of migration through which authoritarian societies passed: migration as an escape from a closed society, in which exit was forbidden and entry encouraged (such as from socialism to capitalism); migration as a two way road in which exit is free and entry welcomed; and lastly, crisis, in which exit is free and entry is closed (Milan Mesic, "External Migration in Post-War Yugoslavia," in John Allcock, John J. Horton and Marko Milivojevic, ed., *Yugoslavia in Transition* [New York: Berg, 1992]).
26. This is contrary to the trend in some Islamic countries, such as Pakistan and Iran, which have extended open arms to the Bosnian Muslims.
27. *The New York Times,* September 26, 1992.
28. *The New York Times,* August 31, 1992.
29. D. N. Chorafas, *The Knowledge Revolution* (New York: McGraw-Hill, 1968), 56.
30. Joly, 83
31. Ibid., 84
32. *The New York Times,* October 14, 1992.
33. There has been much ill feeling among the Serbian authorities because, although the Serbs are housing one-quarter of all the refugees, they receive only 14 percent of the western aid for refugees (*The New York Times,* November 28, 1992).

34. Mark Mazower, *Greece and the Inter-War Economic Crisis* (Oxford: Clarendon Press, 1991), 129-42.
35. *The New York Times,* November 28, 1992. This is putting excessive strain on urban families that are not self-sufficient in food, since according to official Yugoslav statistics, the concentration of refugees is in the major cities of Serbia and adjoining areas (Republicki Zavod Za Statistiku, Republika Srbija, "Izbegla Lica," in *Saopstenje,* no. 11 (January 1993): Table 1).
36. T. W. Shultz estimated losses on the basis of educational costs, Becker, Mincver and Bowman emphasized internal rates of return. For a discussion of these, see Robert Myers, *Education and Immigration* (New York: David McKay Co 1972), 178.
37. Chorafas, 7.
38. *Transition,* vol. 2, no. 2 (February 1991).
39. Joly, 78, 77.
40. Horizont, Bulgarian Radio, January 23, 1991, cited in RFE/RL Research Report, vol. 2, no. 6 (February 5, 1993): 58.
41. RFE/RL News Brief, December 28-January 8, 1993, 14.
42. This is studied in detail by the author in *The Demographic Struggle for Power* (manuscript in preparation).
43. The last census to include an ethnic breakdown was in 1975. The most recent census in 1985 omitted the issue altogether. See RFE/RL Research Report, vol. 2, no. 6 (February 5, 1993): 58-61.
44. *Post-Soviet/East European Report* 9, no. 39 (December 1, 1992), 1.
45. Srdjan Bogosavljevic, "How Many Serbs Are There?" *Vreme,* no. 84 (May 3, 1993).
46. Sergej Flere, "Cognitive Adequacy of Sociological Theories in Explaining Ethnic Antagonism in Yugoslavia," in Kumar Rupesinghe et al., eds., *Ethnicity and Conflict in a Post-Communist World* (New York: St. Martin's Press, 1992), 263.
47. Andrew Bell-Fialkoff, "A Brief History of Ethnic Cleansing," *Foreign Affairs* 72, no. 3 (Summer 1993): 116. Also see Vladimir Dedijer, *The Yugoslav Auschwitz and the Vatican: The Croatian Massacre of the Serbs During World War II* (Buffalo, New York: Prometheus Books, 1992).
48. Joly, 87.
49. Ibid., 86.
50. Bell-Fialkoff, 110.
51. It is interesting to note that despite its frequency across the globe, it was only condemned by the World Court on April 8, 1993.
52. Bell-Fialkoff, 110.
53. These numbers were taken from the Bell-Fialkoff article. However, there has been a heated debate in the course of the late 1980s and early 1990s as to what those numbers actually are. See Ljubo Boban, "Jasenovac and the Manipulation of History," *East European Politics and Societies* 4 (Fall 1990), and Robert Hayden,

"Recounting the Dead: The Rediscovery and Redefinition of Wartime Massacres in Late and Postcommunist Yugoslavia," in Rubie Watson, ed., *Secret Histories: Memory and Opposition Under State Socialism* (Santa Fe, NM: School of American Research Press, 1993).

54. RFE/RL Research Report, vol. 1, no. 45 (November 13, 1992): 77.
55. RFE/RL News Brief, April 13-26, 1993, 12.
56. *The Economist,* 1993.
57. Peter L. Berger, "Preface," in Uri Ra'anan et al., eds., *State and Nation in Multi-ethnic Societies* (Manchester: Manchester University Press, 1991), ix.

Notes to Chapter 7

1. Flora Lewis, "A 19th Century Balkan Solution," *The New York Times,* May 1, 1993.
2. John Zametica, "The Yugoslav Conflict," *Adelphi,* Paper 270 (Summer 1992): 82.
3. With the exception of the Partisans, one of the three warring factions in the Yugoslav civil war that was fought during World War II. The Partisans contained both Serbs and Croats. However, Nora Beloff claims that while that is true, Serbs and Croats fought each other only when they were recruited into the armies of outside powers (Nora Beloff, "An Analysis of Germany's Yugoslav Policy," London, May 13, 1992).
4. Abraham Lincoln, *Letters and Addresses of Abraham Lincoln* (New York: Unit Book Publishing, 1905).
5. *La Repubblica,* August 5, 1992.
6. Speech at the Carnegie Endowment for International Peace, January 8, 1993. In 1970, President Izetbegovic published *The Islamic Declaration,* in which he claims that Muslims cannot live together with other religions and that he will strive to create a Muslim republic in Bosnia-Herzegovina.
7. *The New York Times,* October 9, 1993.
8. *Vreme,* November 9, 1992, 61.
9. It has been claimed in the past year that the atrocities marking the current Yugoslav civil war have scarred the populations so deeply that no future interaction will be possible. I am skeptical, given the evidence of similar scarring (albeit without the benefit of CNN), during other periods of atrocities (notably those perpetrated by the Ottoman forces in the Balkans, or the World War II genocide against the Serbs of Croatia and Bosnia).
10. As Lippmann was largely responsible for the drawing up of many East European boundaries after World War I, so too Cyrus Vance and David Owen may succeed in such an endeavor in the 1990s.
11. This, in effect, becomes the Third Yugoslavia.

12. RFE/RL Daily Report, October 14, 1993, 7.
13. Hans Kohr, "Disunion Now: A Plea for a Society Based Upon Small Autonomous Units," *Commonwealth,* September 26, 1941.
14. Dean Keith Simonton claims that Eastern Europe presently shows the greatest promise because there, like in the city-states of Italy that produced the Renaissance and the small German states that produced Goethe, Hegel and Mozart, there is no cultural homogeneity (*The New York Times,* March 22, 1992).
15. This is in contrast to Czechoslovakia, where a Czechoslovak identity never developed or was officially encouraged.
16. John Allcock, "Introduction," in John Allcock, John J. Horton and Marko Milivojevic, eds., *Yugoslavia in Transition* (New York: Berg, 1992), 3.
17. Amitai Etzioni, "The Evils of Self-Determination," *Foreign Policy* 89 (Winter 1992-93): 33.
18. Robert Hayden, "Constitutional Nationalism in the Formerly Yugoslav Republics," *Slavic Review* 51, no. 4 (Winter 1992).
19. Oszkar Jaszi (1926), cited in Peter Sipos, "National Conflicts and the Democratic Alternative in the Austro-Hungarian Monarchy and Its Successors," in Uri Ra'anan et al., eds., *State and Nation in Multi-ethnic Societies* (Manchester: Manchester University Press, 1991), 105.
20. Nigel Harris, *National Liberation* (London: I. B. Tauris, 1990), 225.
21. William Pfaff, "Invitation to War," *Foreign Affairs* 72, no. 3 (Summer 1993): 101.
22. The economic components of the creation of small states has been addressed by Hobsbawm. He claimed that the idea of a state existing along the lines of its economic interests is not relevant anymore, leading to other parameters in the definition of states. Indeed, Hobsbawm identifies a proliferation of petty nationalisms as societies look for means to define themselves in new economic terms. In the world of interregional economic interdependence, states are no longer aiming for economic self-sufficiency and rely on international markets for the satisfaction of their needs (Eric Hobsbawm, "Some Reflections on the Break-Up of Britain," *New Left Review* 105 [September-October 1977]).
23. Lewis, "A 19th Century Balkan Solution."
24. Joseph S. Roucek, *The Politics of the Balkans* (New York: McGraw-Hill, 1939), 4.
25. Uri Ra'anan et al., eds., *State and Nation in Multi-Ethnic Societies* (Manchester: Manchester University Press, 1991).
26. Indeed, some Bosnian residents are calling for the creation of a fourth republic within Bosnia-Herzegovina "for normal people." This refers to those with a Yugoslav orientation (*The New York Times,* September 8, 1993).
27. Indeed, Lampe and Jackson claim that historically, the Balkan states are "open rather than closed economies, with market traditions that date back as far as those

of command and custom." John Lampe and Marvin Jackson, *Balkan Economic History, 1550-1950* (Bloomington: Indiana University Press, 1982), 13.

28. RFE/RL Research Report, vol. 1, no. 26 (June 26, 1992), 36.
29. *PlanEcon Report,* vol. 8, nos. 14-15 (April 14, 1992).
30. Nicholas V. Giannaris, *Greece and Yugoslavia, an Economic Comparison* (New York: Praeger, 1984), 243.
31. L. S. Stavrianos, *Balkan Federation* (Hamden, Conn.: Archon Books, 1964), 262-63.
32. *Waterplatte,* August 31, 1993 (cited in *Serbian Unity Congress Weekly Bulletin* 27 [September 3, 1993]: 6).
33. *The Wall Street Journal,* November 3, 1992.
34. Conor Cruise O'Brien, "We Enter Bosnia at Our Peril," in *The Independent,* April 23, 1993.
35. Paul Reid, "Behind the Lines with the Neo-Nazi Croat Militia," *The Miami Herald,* May 16, 1993.
36. The Turkish insistence in isolating Serbia has led to the rejection of participation and protest from both Romania and Greece (*La Repubblica,* November 22 and 26, 1993).
37. Turgut Ozal, *La Turkie en Europe* (Paris: Plon, 1988).
38. Quoted in C. M. Turnbull, *A History of Singapore 1819-1988,* 2d ed. (Singapore: Oxford University Press, 1989), 288.
39. *The New York Times,* October 30, 1991.
40. RFE/RL Daily Report, May 12, 1993.
41. *Encyclopedia Britannica,* vol. 25 (Chicago: 1992), 1005.
42. The loosest form of association, based on economic principles, includes simple trade associations: the establishment of free trade areas, customs unions, common markets and ultimately economic unions.
43. Slovenia was not included in this agreement. Oskar Kovac, "Foreign Economic Relations" in Tomislav Popovic, ed., *The Position and Strategy of Serbia in the New European Order* (Belgrade: Institute of Economic Sciences, 1992), 164.
44. *The Economist*, September 18, 1993, 51.
45. *Nin,* August 6, 1993, 17.
46. Robert Kaplan, *Balkan Ghosts* (New York: St. Martin's Press, 1993), 76.
47. Milan Panic, former prime minister of Yugoslavia, has been promoting his idea of a Balkan economic union, in the absence of which all countries of the Balkans are doomed to inviability (RFE/RL Daily Report, June 21, 1993, 6).
48. According to Willer, not only will a new Yugoslavia emerge, but it will have as its capital Zagreb (Interview in *Duga,* October 9-22, 1993, 30).
49. The Balkan federation that is proposed by Cviic excludes Slovenia, Croatia and Bosnia in their Titoist administrative borders (these states, according to Cviic, are

to be a part of a union with Austria and other former imperial states). Such an exclusion is doomed to failure because it divides religions and ethnic groups across borders (namely, Serbs and Muslims) and is thus a constant provocation to war (Christopher Cviic, *Remaking the Balkans* [New York: Council on Foreign Relations Press, 1991], 105).

50. RFE/RL Daily Report, September 7, 1993, 7. Horvat claims that this integration is necessary because the EC is simply not able to absorb the multitude of small, underdeveloped states.
51. *Borba,* July 1, 1993, 3
52. *The New York Times,* October 11, 1993.
53. With respect to Muslim aid, in May 1993, the Saudi Arabian government donated $20 million more in humanitarian aid, bringing the total since the onset of the war to $100 million (RFE/RL News Briefs, vol. 2, no. 19 [April 26-30, 1993]: 9).
54. Indeed, a government-sponsored advertisement states: "An agent is needed, who knows the area, who can and does understand the people, their habits, their idiosyncrasies, who can work within local customs and who can provide an example of political stability, democracy and economic growth. Greece is this agent" (advertisement in *The New York Times,* November 11, 1992, A9-A12).
55. See Milan Skakun, *Balkan, Enigma Bez Resenja* (Belgrade: Naucna Knjiga, 1992).
56. N. Iorga, "Les Conflicts balkaniques," *Le Monde Slave,* February 1926, 171. There are several authors that support this view, and others that deny the existence of a common unifying bond. See Stavrianos, chapter 1, note 7.
57. Michael Ignatieff, "The Shame of Bosnia," *New York Review of Books* 40, no. 9 (May 13, 1993).
58. Interview with Peter Handke published in *Serbia* 7 (January 20, 1992): 21.
59. This was said in 1848 and is cited in Stavrianos, 58.
60. Kaplan, 48.
61. As mentioned in the Introduction, this study does not address itself to the actual political boundaries or structure of these multiethnic units. However, it is based on the belief that Yugoslavia, in its post-World War II configuration, had great merits both on ethnic and economic grounds, and these merits must be considered in formulating new political configurations.
62. Cited in Alfred Pfabigan, "The Political Feasibility of Austro-Marxist Proposals for the Solution of the Nationality Problem of the Danubian Monarchy," in Uri Ra'anan et al., eds., *State and Nation in Multi-Ethnic Societies* (Manchester: Manchester University Press, 1991), 54.
63. Stavrianos, 2.
64. She also suggests including Turkey and possibly Hungary. Lewis, "A 19th Century Balkan Solution."

65. Dusko Doder, "Yugoslavia, New War, Old Hatreds," *Foreign Policy* 91 (Summer 1993): 22.
66. Zametica, 82.

REFERENCES

Official Sources

Comisia Nationala Pentru Statistica. *Romanian Statistical Yearbook.* Bucharest, 1991.

Federal Statistical Office. *Statisticki Godisnjak Jugoslavije 1992.* Belgrade, 1992.

National Statistical Service of Greece. *Statistical Yearbook of Greece 1986.* Athens, 1987.

Savezni Zavod Za Statistiku. *Statisticki Godisnjak Jugoslavije.* Beograd, annual volumes.

The World Bank. *World Development Report 1989, 1990, 1991.* New York: Oxford University Press, 1989, 1990, 1991.

Books and Articles

Barraclough, Geoffrey, ed. *The Times Atlas of World History,* 3d ed. Maplewood, New Jersey: Hammond, 1989.

Bell-Fialkoff, Andrew. "A Brief History of Ethnic Cleansing." *Foreign Affairs* 72, no. 3 (Summer 1993).

Bogosavljevic, Srdjan, ed. *Bosna i Herzegovina Izmedu Rata i Mira.* Belgrade: Dom Omladine, 1992.

Bookman, Milica Zarkovic. *The Economics of Secession.* New York: St. Martin's Press, 1993.

Cohen, Lenard. *Broken Bonds.* Boulder: Westview Press, 1993.

Connor, Walter. "Politics of Ethnonationalism." *Journal of International Affairs* 27, no. 1 (1973).

Cviic, Christopher. *Remaking the Balkans.* New York: Council on Foreign Relations Press, 1991.

Day, Alan J., ed. *Border and Territorial Disputes.* 2d ed. Harlow, Essex: Longman, 1987.

Dedijer, Vladimir. *The Yugoslav Auschwitz and the Vatican.* Buffalo, N.Y.: Prometheus Books, 1992.

Djilas, Aleksa. "Serbia's Milosevic: A Profile." *Foreign Affairs* 72, no. 3 (Summer 1993).

Doder, Dusko. "Yugoslavia, New War, Old Hatreds." *Foreign Policy* 91 (Summer 1993).

Doxey, Margaret P. *Economic Sanctions and International Enforcement.* London: Oxford University Press, 1991.

Etzioni, Amitai. "The Evils of Self-Determination." *Foreign Policy* 89 (Winter 1992-93).

Freris, A. F. *The Greek Economy in the Twentieth Century.* New York: St. Martin's Press, 1986.

Gallagher, Tom. "Vatra Romaneasca and the Resurgent Nationalism in Romania." *Ethnic and Racial Studies* 15, no. 4 (October 1992).

Gati, Charles. "From Sarajevo to Sarajevo." *Foreign Affairs* 71, no. 4 (Fall 1992).

Giannaris, Nicholas V. *Greece and Yugoslavia, An Economic Comparison.* New York: Praeger, 1984.

Glenny, Misha. *The Fall of Yugoslavia.* London: Penguin Books, 1992.

Gordenker, Leon. *Refugess in International Politics.* New York: Columbia University Press, 1987.

Halpern, Morton H., and David J. Scheffer. *Self-Determination in the New World Order.* Washington: Carnegie Endownment for International Peace, 1992.

Hayden, Robert. "Constitutional Nationalism in the Formerly Yugoslav Republics." *Slavic Review* 51, no. 4 (Winter 1992).

Hinic, Branko, and Rajko Bukvic. "Efekti Mera Ekonomske Politike i Ocekivana Privredna Kretanja u Narednom Periodu." Unpublished paper, Belgrade, 1992.

Hufbauer, Gary Clyde, Jeffrey J. Schott and Kimberly Ann Elliot. *Economic Sanctions Reconsidered.* Washington: Institute for International Economics, 1990.

Joly, Daniele. *Refugees.* Boulder: Westview Press, 1992.

Kaplan, Robert. *Balkan Ghosts.* New York: St. Martin's Press, 1993.

Kohr, Hans. "Disunion Now: A Plea for a Society Based Upon Small Autonomous Units." *Commonwealth,* September 26, 1941.

Lampe, John, and Marvin Jackson. *Balkan Economic History, 1550-1950,* Bloomington: Indiana University Press, 1982.

Mazower, Mark. *Greece and the Inter-War Economic Crisis.* Oxford: Clarendon Press, 1991.

Milojic, Andra. "Vrednost Nabavki i Isporuka Izmedu Republika i Pokrajina u 1987 Godini." *Saopstenje* 324 (November 1991).

Neal, Alfred C., ed. *Introduction to War Economics.* Chicago: Richard D. Irwin, 1942.

Pfaff, William. "Invitation To War." *Foreign Affairs* 72, no. 3 (Summer 1993).

Pipinelis, P. *Europe and the Albanian Question.* Chicago: Argonaut, 1963.

Popovic, Tomislav. "Alternativni Scenariji Uredjenja Odnosa Sa Medjunarodnim Okruzenjem i Ocene Njihovih De-Stabilizacionih Efekata." *Osnove Stabilizacionog Programa.* Belgrade: Institut Ekonomskih Nauka, 1992.

———, ed. *The Position and Strategy of Serbia in the New European Order.* Belgrade: Institute of Economic Sciences, 1992.

Poulton, Hugh. *Balkans: Minorities and States in Conflict.* London: Minority Rights Publications, 1991.

Ra'anan, Uri, et al., eds. *State and Nation in Multi-Ethnic Societies,* Manchester: Manchester University Press, 1991.

Ramet, Sabrina P. *Nationalism and Federalism in Yugosalavia,* 2d ed. Bloomington: Indiana University Press, 1992.

Ramet, Sabrina Petra. "War in the Balkans." *Foreign Affairs* 71, no. 4 (Fall 1992).

Roucek, Joseph S. *The Politics of the Balkans.* New York: McGraw-Hill, 1939.

Rudolph, Joseph R., Jr., and Robert J. Thompson, eds. *Ethnoterritorial Politics, Policy, and the Western World.* Boulder: Lynne Rienner Publishers, 1989.

Rudolph, Richard L., and David Good, eds. *Nationalism and Empire.* New York: St. Martin's Press, 1992.

Rupesinghe, Kumar, Peter King and Olga Vorkunova, eds. *Ethnicity and Conflict in a Post-Communist World.* London: Macmillan Press, 1992.

Rusinow, Dennisow. "Yugoslavia: Balkan Breakup?" *Foreign Policy* 83 (Summer 1991).

Sachs, Jeffrey. "Poland and Eastern Europe: What Is to Be Done?" in Andras Koves and Paul Marer, eds. *Foreign Economic Liberalization.* Boulder: Westview Press, 1991.

Samarasinghe, S. W. R. de A., and Reed Coughlan, eds. *Economic Dimensions of Ethnic Conflict.* London: Pinter Publishers, 1991.

Samardzic, Radovan, and Dimitrije Dordevic. *Migrations in Balkan History.* Belgrade: Institute for Balkan Studies Special Edition no. 39, 1989.

Shiels, Frederick L., ed. *Ethnic Separatism and World Politics,* Lanham, Md.: University Press of America, 1984.

Sjoberg, Orjan, and Michael L. Wyzan. *Economic Change in the Balkan States.* New York: St. Martin's Press, 1991.

Skakun, Milan. *Balkan, Enigma Bez Resenja.* Belgrade: Naucna Knjiga, 1992.

Smith, Anthony. "Chosen Peoples: Why Ethnic Groups Survive." *Ethnic and Racial Studies* 15, no. 3 (July 1992).

Stavrianos, L. S. *Balkan Federation.* Hamden, Conn.: Archon Books, 1964.

Zametica, John. "The Yugoslav Conflict." *Adelphi.* Paper 270 (Summer 1992).

Zaslavsky, Victor. "Nationalism and Democratic Transition in Postcommunist Societies." *Daedalus* 121, no. 2 (Spring 1992).

INDEX